144

Advances in Polymer Science

Springer-Verlag Berlin Heidelberg GmbH

Polymer Latexes
Epoxide Resins
Polyampholytes

With contributions by
M. Ballauff, A.E. Batog, J. Bolze,
N. Dingenouts, S.E. Kudaibergenov,
P. Penczek, I.P. Pet'ko, D. Pötschke

 Springer

This series presents critical reviews of the present and future trends in polymer and biopolymer science including chemistry, physical chemistry, physics and materials science. It is addressed to all scientists at universities and in industry who wish to keep abreast of advances in the topics covered.

As a rule, contributions are specially commissioned. The editors and publishers will, however, always be pleased to receive suggestions and supplementary information. Papers are accepted for „Advances in Polymer Science" in English.

In references Advances in Polymer Science is abbreviated Adv. Polym. Sci. and is cited as a journal.

Springer WWW home page: http://www.springer.de

ISSN 0065-3195

ISBN 978-3-662-14720-7   ISBN 978-3-540-68384-1 (eBook)
DOI 10.1007/978-3-540-68384-1

Library of Congress Catalog Card Number 61642

© Springer-Verlag Berlin Heidelberg 1999
Originally published by Springer-Verlag Berlin Heidelberg New York in 1999
Softcover reprint of the hardcover 1st edition 1999
The use of registered names, trademarks, etc. in this publication does not imply, even in the absence of a specific statement, that such names are exempt from the relevant protective laws and regulations and therefore free for general use.

Typesetting: Data conversion by MEDIO, Berlin
Cover: E. Kirchner, Heidelberg
SPIN: 10648266   02/3020 - 5 4 3 2 1 0 - Printed on acid-free paper

Editorial Board

Contents

Analysis of Polymer Latexes by Small-Angle X-Ray Scattering

N. Dingenouts[1], J. Bolze[1], D. Pötschke[1], M. Ballauff [1,2]

[1] Polymer-Institut, Universität Karlsruhe, Kaiserstrasse 12, 76128 Karlsruhe, Germany
[2] e-mail: matthias.ballauff@chemie.uni-karlsruhe.de

This article reviews the structural analysis of polymeric latexes by small-angle X-ray scattering (SAXS). It is demonstrated that SAXS is a tool which allows us to study the internal structure as well as the interaction of latex particles with great accuracy. Since the SAXS-analysis of latexes requires highly accurate data down to smallest scattering angles, a detailed description of the experimental procedure and the subsequent data treatment will be given. In addition, a brief survey of instruments used up to now for SAXS-studies on latexes will be presented. Because of the low electron density of most of the polymers used in emulsion polymerization SAXS may be applied in conjunction with contrast variation. This method can be used for a precise analysis of the radial structure of the latex particles. Its application to the structural analysis of core-shell particles, to swollen latexes and to surfactants adsorbed on the particles surface will be discussed in detail.

Keywords: Latex, Small-angle X-ray scattering, Small-angle neutron scattering, Wall-repulsion effect, Surfaces

List of Symbols and Abbreviations

Abbreviations

MMA	methylmethacrylate
PMMA	poly(methylmethacrylate)
PS	poly(styrene)
SANS	small-angle neutron scattering
SAS	small angle scattering
SAXS	small-angle X-ray scattering
SDS	sodium dodecylsulfate
USAXS	ultrasmall-angle X-ray scattering

Symbols

α	parameter characterizing the dependence of R_g on contrast	Eq.(21)
β	parameter characterizing the dependence of R_g on contrast	Eq.(22)
b_n	scattering length density (SANS)	Eq.(29)
$B_0(q)$	scattering amplitude of a homogeneous sphere	Eq.(12)
D_n	number-average diameter of latex spheres	
$\varepsilon(q)$	scattering amplitude calculated from the variation of electron density inside of the sphere	Eq.(13), (15)
$I(q)$	scattered intensity	Eq.(1)
I_{fluct}	scattering intensity due to fluctuations of electron density	
$I_0(q)$	scattering intensity of a single particle	Eq.(2)
$I_S(q)$, $I_{SI}(q)$, $I_I(q)$	partial scattering intensities	Eq.(8)
$\bar{I}(m)$	smeared intensity	Eq.(31)
$\hat{I}(m)$	smeared intensity	Eq.(32)
m	coordinate of position-sensitive detector	Fig. 9
N_i	number density of particles of species i	
q	magnitude of scattering vector; $q=(4\pi/\lambda)\sin(\theta/2)$; λ: wavelength of radiation, θ: scattering angle; cf. Refs.[1–5]	
$\rho(\vec{r})$	local electron density	Eq.(3)
$\rho_p(\vec{r})$	local electron density of particle	
ρ_m	electron density of medium (serum)	
$\bar{\rho}$	average electron density of particle	Eq.(5)
$\Delta\rho$	parameter characterizing invariant Q	Eq.(26)

$\overline{\Delta\rho^2}$	parameter characterizing invariant Q	Eq.(27)
$\Delta\rho_i, \Delta\rho_i$	difference in electron density of adjacent phases	Eq.(28)
Q	invariant	Eq.(23)
R	radius of spheres	
R_g	radius of gyration	Eq.(19)
$R_{g,\infty}$	radius of gyration of shape function	Eq.(20)
r_o	Thomson radius of the electron; scattering length of a single electron	
$S(q)$	structure factor	Eq.(1)
S_i, S_a	magnitude of inner and outer surface of particle, respectively	Eq.(28)
$T(\bar{r})$	shape function	Eq.(3)
V_p	volume of particle	Eq.(4)
V_c	volume defined according to	Eq.(25)

1
Introduction

Small-angle X-ray scattering (SAXS) has become a well-established tool in colloidal science and has been applied to a great number of polymeric systems and colloids [1–5]. A survey of the vast literature pertinent to investigations by SAXS shows, however, that this method had only rarely been applied to latexes until recently. On the other hand, the analysis of polystyrene latexes with narrow size distribution [6,7] played an important role in the early history of this method. A number of early workers in this field [8–12] showed that the minima and maxima seen in the scattering curve of a polystyrene latex (see the discussion on p. 54 of Ref. [1]) are related to the form factor of a homogenous sphere and may be used to determine the size of the particles. A review of the older investigations which all assumed a homogeneous electron density distribution within the spheres was given some time ago by Vanderhoff [13].

Small-angle neutron scattering (SANS) [14,], on the other hand, has been used quite frequently to analyze the radial structure [15–27] and the surface [28–30] of latex particles. The reason for the great number of applications of SANS can be traced back to the fact that a high contrast between the particle and the medium or between the different constituents of the particle may be achieved through appropriate substitution of hydrogen atoms by deuterium atoms [14]. In particular, a variable contrast between the latex particles and the surrounding medium water can be adjusted through mixtures of H_2O and D_2O [14]. This allows the detailed study of the radial structure and surface of the particles through contrast variation which had been earlier established as an investigative tool for polymeric systems [31–37], biological structures (see Refs. [38,39] and further citations given there), and in general for colloidal systems [40,41].

Another reason for the frequent application of SANS to latexes is located in the fact that latex particles often exhibit a radius of 100 nm and more. This requires very small scattering angles for a meaningful analysis by small-angle scattering, i.e, the magnitude of the scattering vector q ($q=(4\pi/\lambda)\sin(q/2)$; λ: wavelength of radiation, θ: scattering angle; [1–5]) must attain values much smaller than 0.1 nm^{-1} to enable a full analysis of the data to be made. The exceedingly small scattering angles thus necessitated impose no particular problem on a neutron spectrometer as e.g. the D11 at the Institut Laue-Langevin [42] but are difficult to achieve with conventional SAXS equipment.

In the course of a number of recent studies [43–55], it has been demonstrated that SAXS is an excellent tool for the study of polymeric latexes. Given the possibility of performing highly precise measurements extending down to considerably lower q-values ($q\geq0.025$nm^{-1}) SAXS can now be used for investigating latex particles up to a diameter of 200 nm. The electron densities of the polymers commonly used for the synthesis of such particles differ markedly thus allowing the study of composite particles. Moreover, it has been recognized [47] that the low electron densities of polymers as e.g. polystyrene (PS) or poly(methyl methacrylate) (PMMA) allow to reach the match point of typical latex particles. Thus, by adding sucrose or glycerol to the serum of the latex the electron density of the suspending medium may be raised sufficiently to match the electron density of latex-particles composed of vinyl polymers. Another advantage is given by the fact that SAXS has no incoherent background [3, 14] and the scattering intensities of the latexes can be measured up to high scattering angles.

In the present article, we review recent SAXS-studies conducted on polymeric latexes. We will first give an exposition on the theory of SAXS including contrast variation based on references [56–60]. The main purpose of the theoretical exposition is a clear assessment of the structural information embodied in the SAXS-intensities. This discussion will also be helpful to delineate possible limitations of this method. It will reveal which parameters can be gained from a SAXS-analysis of latex particles and their relation to the structure of the particles.

Since a meaningful analysis of latex particles requires measurements down to smallest q-values, a detailed description of an optimized small-angle camera together with the discussion of experimental problems of SAXS will be given. Also, we shall discuss the subsequent treatment of data and the steps necessary to extract the structural information from the SAXS-intensities. Section 4 is devoted to a discussion of the experimental results obtained in recent investigations.

2
Theory

2.1
Structure of Particles; Contrast Variation

We consider a system of monodisperse particles of number density N which are oriented at random. The scattering intensity $I(q)$ as function of q, the magnitude of the scattering vector is given by [1–5]

$$I(q) = N I_0(q) S(q) \tag{1}$$

where $I_0(q)$ is the scattering intensity of a single particle at infinite dilution and $S(q)$ denotes the structure factor [1,2,5]. Eq.(1) is strictly valid only for a system of spherical particles which is the case for the latex particles under consideration here. The alterations effected by $S(q)$ can be removed by proper extrapolation to vanishing concentration. Furthermore, its influence is restricted to small scattering angles beyond which it can be neglected as will be shown below. In the following we will therefore deal first with $I_0(q)$ which embodies the internal structure of the particles.

For a particle of arbitrary shape immersed in a solvent of scattering length density ρ_m, $I_0(q)$ can in general be calculated through resort to the Debye-equation [1]

$$I_o(q) = \iint [\rho(\vec{r}_1) - \rho_m][\rho(\vec{r}_2) - \rho_m] \frac{\sin(q|\vec{r}_1 - \vec{r}_2|)}{q|\vec{r}_1 - \vec{r}_2|} \, d\vec{r}_1 d\vec{r}_2 \tag{2}$$

where $\rho(\vec{r})$ denotes the local scattering length density. In the case of SAXS, $\rho(\vec{r})$ is the local electron density multiplied by the Thomson factor r_o. Because r_o is a constant, it is often omitted and the SAXS scattering intensity is expressed in units of the scattering intensity of a single electron [1, 3]. Because of the small angle, the intensity of single atoms is simply given by the square of its number of electrons [1–3]. In the case of SANS [14], however, the scattering length of different elements differ widely and must be taken into account (see Sect. 2.3). For a comparison of SAXS and SANS it is therefore expedient to multiply the electron density by the Thomson factor which leads to calculated SAXS-intensities directly comparable with the respective SANS result. Since this review is mainly devoted to SAXS, $I(q)$ is given in units of the scattering intensity of a single electron (e.u.) as outlined above.

It is expedient to render $\rho(\vec{r})$ as the product of a shape function $T(\vec{r})$ and the local electron density $\rho_p(\vec{r})$ inside the particle. For objects with sharp surfaces $\vec{r}$ may assume only the value 1 or 0 depending on whether $T(\vec{r})$ is a point inside or outside of the particle. This condition may also be relaxed and $T(\vec{r})$ is allowed to vary continuously between 0 and 1 [56, 57, 60]. Thus we have

$$\rho(\vec{r}) = T(\vec{r})\rho_p(\vec{r}) + \rho_m(1 - T(\vec{r})) \tag{3}$$

Given this definition of the shape function, the volume V_p of the particle follows as

$$V_p = \int T(\vec{r}) d\vec{r} \tag{4}$$

and its average electron density $\bar{\rho}$ by

$$\bar{\rho} = \frac{1}{V_p} \int T(\vec{r}) \rho_p(\vec{r}) d\vec{r} \tag{5}$$

In the following, the difference $\bar{\rho} - \rho_m$ will be designated as contrast. The local excess electron density may be split into a part depending on the contrast $\bar{\rho} - \rho_m$ and into a function $\Delta\rho(\vec{r})$ independent of contrast:

$$\rho(\vec{r}) - \rho_m = T(\vec{r})[\bar{\rho} - \rho_m] + T(\vec{r})\Delta\rho(\vec{r}) \tag{6}$$

From the definition of the function $\Delta\rho(\vec{r})$ given in Eq.(6) it is evident that

$$\int T(\vec{r})\Delta\rho(\vec{r}) d\vec{r} = 0 \tag{7}$$

Introduction of Eq.(6) into (2) leads to the splitting of the scattering intensity $I_o(q)$ into three parts:

$$I_o(q) = I_S(q) + 2I_{SI}(q) + I_I(q) \tag{8}$$

where

$$I_S(q) = [\bar{\rho} - \rho_m]^2 \iint T(\vec{r}_1)T(\vec{r}_2) \frac{\sin(q|\vec{r}_1 - \vec{r}_2|)}{q|\vec{r}_1 - \vec{r}_2|} d\vec{r}_1 d\vec{r}_2 \tag{9}$$

and

$$I_{SI}(q) = [\bar{\rho} - \rho_m] \iint T(\vec{r}_1)T(\vec{r}_2)\Delta\rho(\vec{r}_2) \frac{\sin(q|\vec{r}_1 - \vec{r}_2|)}{q|\vec{r}_1 - \vec{r}_2|} d\vec{r}_1 d\vec{r}_2 \tag{10}$$

and

$$I_I(q) = \iint T(\vec{r}_1)\Delta\rho(\vec{r}_1)T(\vec{r}_2)\Delta\rho(\vec{r}_2) \frac{\sin(q|\vec{r}_1 - \vec{r}_2|)}{q|\vec{r}_1 - \vec{r}_2|} d\vec{r}_1 d\vec{r}_2 \tag{11}$$

Hence, Eq. (8) to (11) demonstrate that the immersion of the particles in a medium having the electron density ρ_m leads to three different components of the measured intensity which depend differently on the contrast $\bar{\rho} - \rho_m$. The first term $I_S(q)$ presents the Fourier-transform of the shape function $T(\vec{r})$ and its careful determination therefore allows us to deduce all the information solely due to the shape of the particle. The third term, on the other hand, which dominates the measured scattering function at low contrast is related to the interference due to the internal variation of the electron density.

Thus, depending on contrast the same composite particles may exhibit a totally different scattering function. Precise knowledge of $\bar{\rho} - \rho_m$ is therefore necessary for a meaningful assessment of the information embodied in $I_0(q)$. Measurements at three different contrasts at least thus may serve for determining the scattering functions $I_S(q)$, $I_I(q)$ and the cross term $I_{SI}(q)$ separately. This method which has its origins mainly in the work of Kirste and Stuhrmann [34–36] has been termed *contrast variation* and has already been applied to the analysis of colloidal particles [40, 41, 47, 51, 58].

Isoscattering point. In particular, Eq.(11) shows that $I_I(q)$ is independent of contrast. If the part $I_S(q)$ being determined by the shape of the particles as well as the cross term $I_{SI}(q)$ exhibit deep minima for certain values of q*, i.e., all scattering curves must cross at q* regardless of contrast. These "isoscattering points" have been discussed by a number of authors [40, 58, 59, 60, 61], in particular for nearly spherical objects. From Eq. (8) to (11) it is clear, however, that the only requirement for the occurrence of crossing points are deep minima of the form part $I_S(q)$ (Eq.(9)) and the cross term $I_{SI}(q)$ (Eq.10)); i.e., the Fourier-transform of the shape function must exhibit pronounced minima or roots.

For spheres having an outer radius R both terms (9) and (10) may be simplified considerably. In this case the scattering intensity may be rendered as [40, 47]

$$I_0(q) = B^2(q) \tag{12}$$

with

$$B(q) = B_0(q) + \varepsilon(q) \tag{13}$$

where

$$B_0(q) = (\bar{\rho} - \rho_m) 4\pi \int_0^R T(r) \frac{\sin(qr)}{qr} r^2 dr \quad . \tag{14}$$

and

$$\varepsilon(q) = 4\pi \int_0^R T(r) \Delta\rho(r) \frac{\sin(qr)}{qr} r^2 dr \tag{15}$$

The quantity $B_0(q)$ presents the scattering amplitude of a homogeneous sphere whereas $\varepsilon(q)$ solely refers to the variation of ρ inside the sphere. $B_0(q)$ will vanish for tan(q*R)=q*R and $I_0(q*)=\varepsilon^2(q*)$. Hence, in the case of well-defined particles with spherical symmetry the isoscattering points present a prominent feature of the scattering curves as function of contrast and may be used to determine R.

In the following the isoscattering point shall be discussed for a monodisperse core-shell sphere. The radial electron density profile is displayed in Fig. 1. There is a shell of three nanometers thickness in which the electron density is increased by 20 electrons /nm^3.

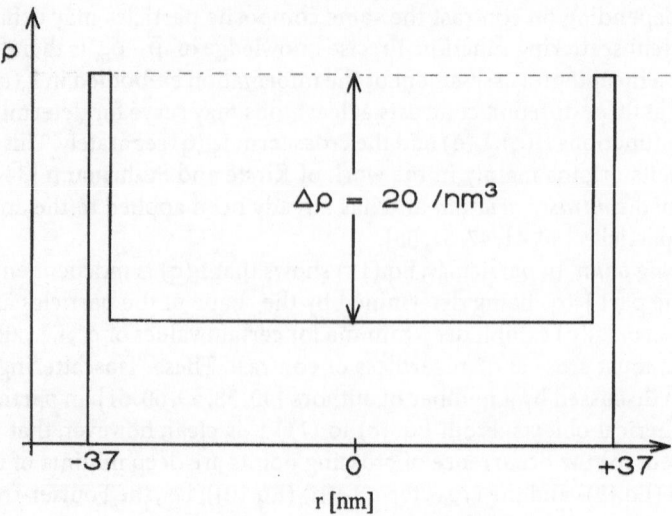

Fig. 1. Radial electron density used in the model calculation (see Figs. 2, 4 and 5)

Figure 2 displays the SAXS-intensities $I_0(q)$ calculated for the radial electron density shown in Fig. 1. Parameter of the curves is the contrast $\bar{\rho} - \rho_m$ expressed as the number of excess electrons per nm^3. The isoscattering points are clearly visible. Furthermore, the calculation shows that forward scattering for a monodisperse particle will vanish at zero contrast in accordance with the above deductions. As a consequence of this, the radius of gyration will increase or decrease rapidly as function of contrast in the vicinity of the match point (see below).

Figure 2 also shows that the isoscattering points are features of $I_0(q)$ which are located in a q-range easily accessible by conventional SAXS-equipment ($q > 0.08 nm^{-1}$). A careful study of composite particles by contrast variation may therefore serve to elucidate their internal structure without the necessity to explore the region of smallest q-values (see below the discussion of the radius of gyration).

Effect of size polydispersity. For vanishing concentration, Eq. (1) can be generalized to polydisperse systems by adding up the contributions $I_{0,i}(q)$ of all species i weighted by their respective number density N_i:

$$I_0(q) = \sum_i N_i I_{0,i}(q) \tag{16}$$

The effect of polydispersity can be seen directly when considering first a system of homogeneous spheres having a Gaussian size distribution [46]. Here the scattering curves are fully determined by the form amplitude $B_0(q)$ as defined

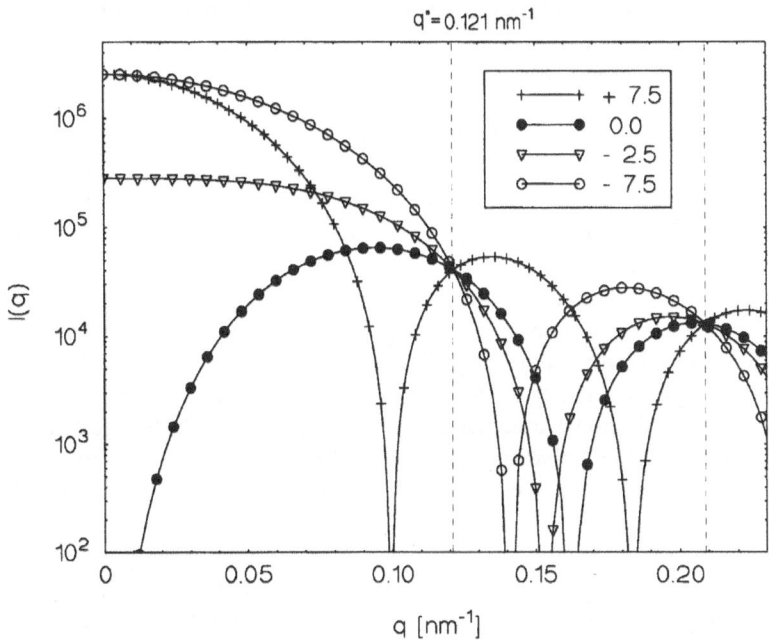

Fig. 2. Isoscattering points for a monodisperse core-shell sphere. The electron density is displayed in Fig. 1. The inset gives the contrast $\bar{\rho} - \rho_{\text{m}}$ (electrons/nm³)

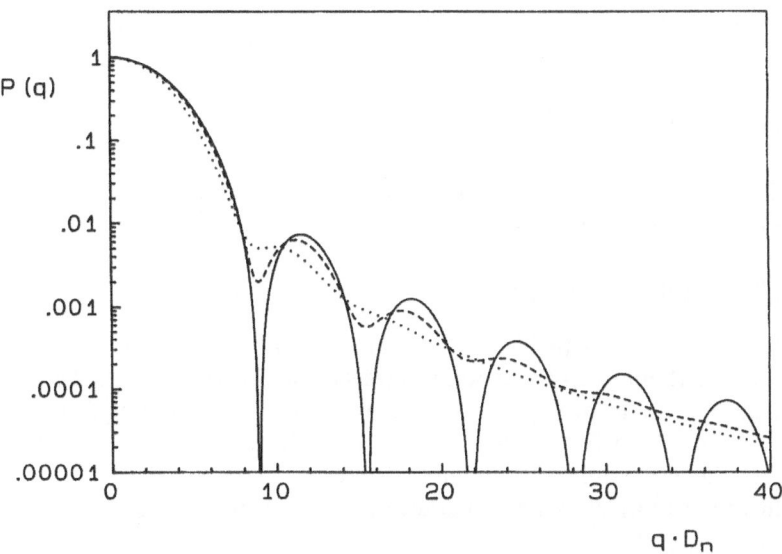

Fig. 3. Form factor P(q) of homogeneous spheres with a Gaussian size distribution at constant number average diameter D_n calculated for different standard deviations σ. *Solid line*: σ=0 nm; *dashed line*: σ/D_n=7.5%; *dotted line*: σ/D_n =15% (taken from Ref. [46])

through Eq.(14). Fig. 3 displays the form factor P(q) which is given by the ratio of $I_0(q)$ to $I_0(0)$. The calculation has been done for three different non-uniformities as expressed by their respective standard deviations. The abscissa has been scaled by the number average diameter of the spheres.

Polydispersity has a profound influence because it smears out the deep minima or zeros of the form factor. For standard deviations above 15% the SAXS-analysis of latex particles becomes very difficult because in these cases the minima of P(q) have nearly disappeared. The same holds true for the SANS-analysis of such systems, of course.

In the case of inhomogeneous particles polydispersity will obscure the isoscattering points too because these features are located directly at the q-values of minima of the form part $I_S(q)$. Therefore the crossing points will become apparent only in the vicinity of the match point where the influence of $I_S(q)$ is small. This can be demonstrated when considering the same system as discussed in Fig.1 but now with a polydisperse size distribution which was assumed to be Gaussian with a standard deviation of 9%. The result of this calculation displayed in Fig. 4 refers directly to the scattering intensities of latex particles exhibiting a core-shell structure [47, 48, 52]. Figure 4 points to the importance of contrast variation: It shows that only the first isoscattering point is clearly visible; at higher q this feature is no longer discernible except for low contrast. Polydispersity may even hide the isoscattering points for spheres having only a thin shell differing in electron density from the core. In this case it is necessary to conduct the measurements in the immediate neighborhood of the match point as suggested by the model calculation shown in Fig. 4 [52].

Contrast variation; low-angle part of I(q). Series expansion of Eq.(2) leads to Guinier's law [1]:

$$I_0(q) \cong I_0(0)\exp[-\frac{1}{3}R_g^2 q^2] \tag{17}$$

$I_0(0)$ and the radius of gyration R_g are now related to the local excess electron density and it is easy to show that

$$I_0(0) = I_S(0) = V_P^2[\bar{\rho} - \rho_m]^2 \tag{18}$$

Therefore plots of $I_0(0)^{1/2}$ vs. ρ_m can be used to determine the average-electron density. It must be kept in mind, however, that Eq.(18) assumes that all particles have the same average electron density, i.e, there is no polydispersity of contrast. Otherwise the particles will exhibit a finite forward scattering intensity $I_0(0)$ (cf. the discussion of this point below).

The radius of gyration has a marked dependence on contrast through [3, 36, 38]

$$R_g^2 = R_{g,\infty}^2 + \frac{\alpha}{\varrho - \varrho_m} - \frac{\beta}{(\varrho - \varrho_m)^2} \tag{19}$$

where

$$R_{g,\infty}^2 = \frac{1}{V_p}\int T(\vec{r})r^2 d\vec{r} \tag{20}$$

$$\alpha = \frac{1}{V_p}\int T(\vec{r})\Delta\rho(\vec{r})r^2 d\vec{r} \tag{21}$$

$$\beta = \frac{1}{V_p^2}\left[\int T(\vec{r})\Delta\rho(\vec{r})\vec{r}\,d\vec{r}\right]^2 \tag{22}$$

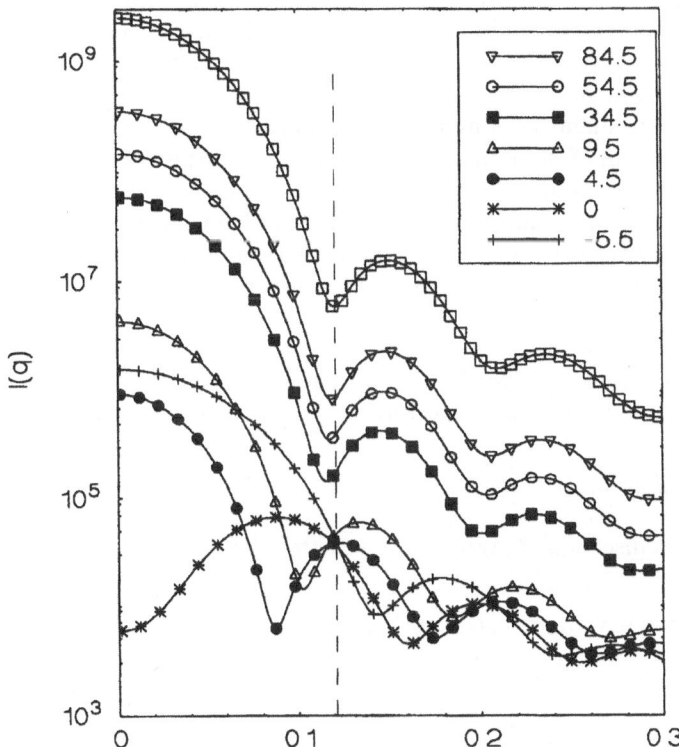

Fig. 4. Isoscattering point for a system of polydisperse core-shell spheres. A Gaussian size distribution with a standard deviation of 9% has been assumed for the cores whereas the thickness of the shell was kept constant. The inset gives the contrast $\bar{\rho} - \rho_m$ (electrons/nm³). The *uppermost curve* refers to homogeneous spheres with diameter 37 nm. The *dashed line* marks the isoscattering point which coincides with the minimum of the form factor of the homogeneous sphere

$R_{g,\infty}$ is the radius of gyration of the shape function whereas the quantity α is related to the internal distribution of the electron density. The quantity β is the square of the distance between the centers of gravity of $T(\vec{r})$ and of the internal distribution $T(\vec{r})\Delta\rho(\vec{r})$ and vanishes for particles with centrosymmetric structure.

For a homogeneous sphere with radius R, $R_g = R_{g,\infty}$ which may be calculated by the relation $R_g^2 = 3/5\ R^2$. Thus, a homogenous sphere of 100 nm diameter is characterized by $R_g = 38.73$ nm. The validity of the Guinier-law Equation (17) requires that $R_g q < 1$ and q must therefore be much smaller than 0.1 nm^{-1}. Often, this prerequisite for the application of Eq.(17) is not fulfilled and the radius of gyration cannot be measured for latex systems with sufficient accuracy.

Equation(19) demonstrates that R_g of a composite particle diverges at the match point and that R_g^2 may become negative as well. This is shown in Fig. 5 for concentric monodisperse core-shell particles the radial electron density of which is given by Fig. 1 whereas the scattering function has already been discussed in conjunction with Fig. 2. Polydispersity of contrast tends to smear out this feature but it should be kept in mind that R_g^2 may change markedly when conducting measurements in the immediate vicinity of the match point.

As already mentioned above, the radii of gyration of homogeneous latex particles having a diameter of more than 100 nm are too large to be determined by conventional SAXS-equipment. In the case of composite particles this problem may even be aggravated by the rise of R_g in the vicinity of the match point. It is thus clear that up to now the SAXS-analysis of latex particles could not attain this quantity due to the restricted q-range. Very often, however, the SAXS-intensities measured at higher q-values contain enough information to allow an analysis of the radial electron density of the latex particles. This point has already been mentioned when discussing the isoscattering point.

Contrast variation; invariant. The invariant Q presents the integral over the entire scattering intensity in the reciprocal space and depends on contrast as well:

$$Q = \int I(q)q^2 dq = 2\pi^2 \int [\rho(\vec{r}) - \rho_m]^2 d\vec{r} \qquad (23)$$

Through insertion of (6) into (23) we obtain [60]

$$Q = 2\pi^2 V_c [(\overline{\rho} - \rho_m)^2 + 2(\overline{\rho} - \rho_m)\overline{\Delta\rho} + \overline{\Delta\rho^2}] \qquad (24)$$

where

$$V_c = \int T^2(\vec{r})d\vec{r} \qquad (25)$$

and

$$\overline{\Delta\rho} = \frac{1}{V_c} \int T^2(\vec{r})\Delta\rho(\vec{r})d\vec{r} \qquad (26)$$

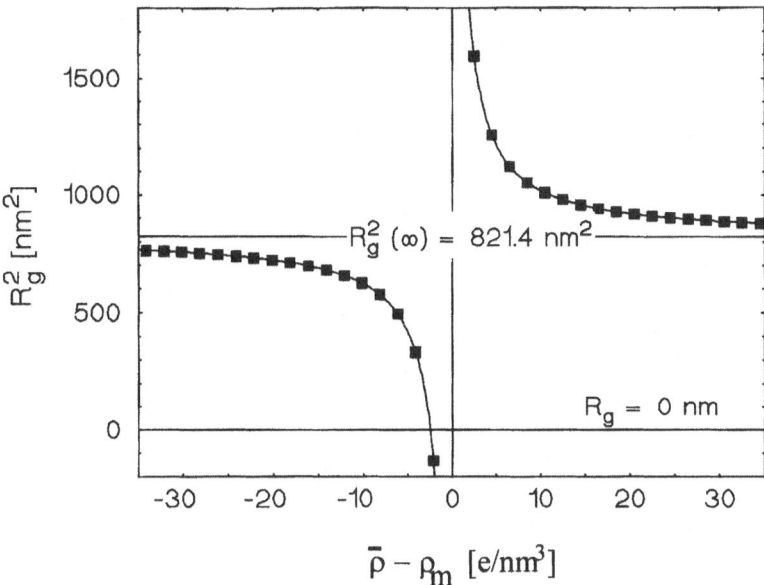

Fig. 5. Radius of gyration of monodisperse core-shell particles as function of contrast. The electron density profile shown in Fig. 1 has been used for the calculation. The plot demonstrates the divergence of R_g^2 in the vicinity of the match point (cf. Eq. (19))

$$\overline{\Delta\rho^2} = \frac{1}{V_c}\int T^2(\vec{r})\Delta\rho^2(\vec{r})d\vec{r} \qquad (27)$$

Here it has been assumed that the shape function $T(\vec{r})$ may vary continuously between 1 and 0 [57, 60]. From Eq.(7) and (26) it is obvious that the coefficient $\overline{\Delta\rho}$ will vanish for particles characterized by a discontinuous shape function $T(\vec{r})$ which is restricted to assume either 1 or 0. A direct consequence of a continuous shape function is furthermore given by the fact that the volume V_c is in general smaller than V_p since Eq.(25) presents the volume integral over $T^2(\vec{r})$ where $T(\vec{r})$ is in general smaller than unity. Only for particles with sharp boundaries the resulting V_c is equal to V_p defined as the integral over $T(\vec{r})$ (Eq.(4)).

It is thus evident that V_c can be distinctly different from V_p. The latter quantity derives from the forward scattering of the particles and is truly invariant against $T(\vec{r})$. The invariant, however, is governed by the scattering intensities at high q where a continuous shape function will lead to profound alterations of $I_0(q)$.

The invariant has played a major role in the development of theory [1–3] and it is often recommended that Q be used to avoid the measurements of I(q) on an absolute scale. For homogeneous particles, this is directly obvious since here $V_c=V_p$ and the parameters defined in Eq.(26) and (27) vanish. Hence, Q= $2\pi^2 V_p(\bar{\rho}-\rho_m)^2$ and division of I(q) bQ (see Eq.(17) and (18)) leads to a quantity independent of contrast and thus obviates the need of absolute scattering intensities (see the discussion of this point in Sect. 3). The above discussion of Q has shown, however, that the invariant consists in general of three terms of which the first one (Eq.(24)) is only the leading one if the contrast is high. In consequence, at intermediate contrast Q is not suitable to normalize the data. In addition to this, the calculation of Q which requires very accurate data at high q-values is often difficult because of the fluctuation-induced contribution to the scattering intensity must be subtracted accurately.

Final slope. For high q the scattering intensity of composite latexes follows Porod's law [1–3]

$$I(q) \rightarrow 2\pi N[\Delta\rho_i^2 S_i + \Delta\rho_a^2 S_a]q^{-4} + I_{fluct} \tag{28}$$

Here S_i and S_a denote the inner interface and the outer surface of the particles, respectively, and the $\Delta\rho$ are the differences in scattering electron density of the adjacent phases. The contribution I_{fluct} is the scattering intensity due to density fluctuations in the solid polymers [2, 14, 47, 85].

It must be kept in mind that the application of Eq.(28) to the analysis of latex systems requires a subtraction of the scattering contribution of the serum with high precision, otherwise the error in the region of high scattering angles becomes too large for a determination of S with sufficient precision. Also, I_{fluct} depends on q at high scattering angles which must be taken into account accordingly [47]. If the diameter of the latex spheres is quite high as e.g. of the order of 300–400 nm, however, the q-range in which Eq.(28) is valid is shifted to smaller scattering angles in which the problems of a fluctuation-induced scattering I_{fluct} or the scattering by the serum are of minor importance. In this case Eq.(28) may be applied directly without problems. It thus may provide a means to estimate the magnitude of S_i if S_a, $\Delta\rho_a$ and $\Delta\rho_i$ are known with sufficient accuracy.

2.2
Interaction of Particles; Structure Factor

As mentioned above, measurements at finite concentrations lead to a non-vanishing influence of the structure factor S(q). For the overwhelming majority of the latex systems studied by SAS-experiments so far, colloid stability has been achieved by a screened Coulomb interaction [5, 62, 63]. The structure factor of such a system of particles interacting through a Yukawa-potential has been extensively studied theoretically by Klein and coworkers (see Ref. [63] and further citations given there) who extended the treatment to polydisperse systems.

In principle, SAXS is suitable for determining the structure factor S(q) from experimental results. It will become apparent, however, that the influence of particle interaction in suspensions of latex particles is restricted to very small q-values. This is due to the size of typical latex particles which is of the order of 100 nm and more which requires measurements far below 0.1 nm^{-1}. This can be discussed best in terms of model calculations.

Figure 6 shows the structure factor S(q) calculated for electrostatic repulsion (dotted line; dashed line) and for hard core repulsion (solid line) for a system of spheres having a diameter of 80 nm and a volume fraction of 20% [46]. Again the abscissa has been scaled by the number-average diameter D_n. At low ionic strength there is a strong electrostatic repulsion between the spheres leading to a pronounced maximum in the structure factor (Fig. 6; dotted line). If the ionic strength in increased, however, the repulsive electrostatic interaction is screened and the variation of S(q) is much weaker (dashed line in Fig. 6). At high ionic strength the electrostatic repulsion of the latex particles is screened considerably and in first approximation the structure factor may be calculated in terms of effective hard sphere interaction (cf. Refs. [64] and [65]).

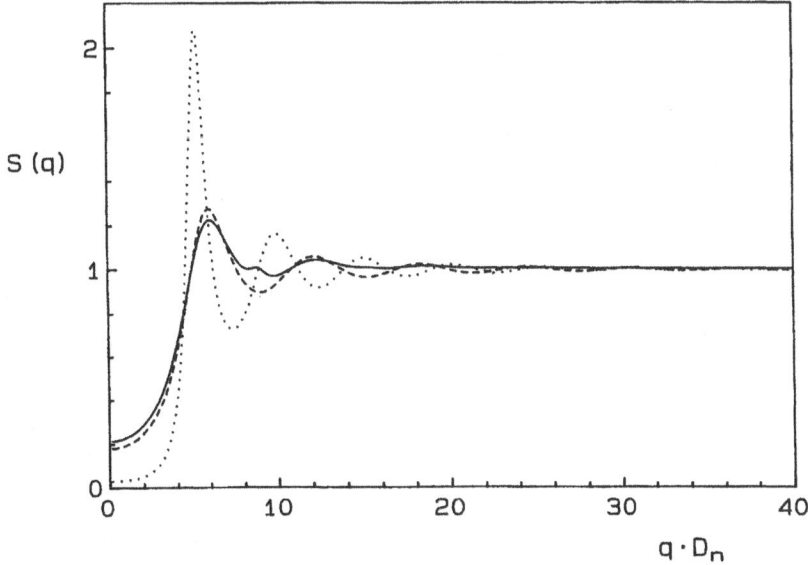

Fig. 6. Structure factor S(q) (cf. Eq.(1)) calculated for a system of homogeneous spheres (volume fraction: 20%; number-average diameter: 80 nm) with hard core interaction (Gaussian size distribution with 7.5% standard deviation; solid line) and with electrostatic interaction [63] (monodisperse, surface charge: 300 e$^-$) at different ionic strength. The *dashed line* refers to high ionic strength (0.02) and the dotted line gives the result obtained for low ionic strength. The strong oscillations of S(q) at low ionic strength would greatly disturb the analysis of $I_0(q)$ (see Eq. (1)) but raising the ionic strength removes the marked influence of S(q) which nearly coincides with S(q) obtained for hard spheres. The data have been taken from Ref.[46]

Polydispersity plays an important role for S(q) as well [63]. Here a discussion of the alterations effected to S(q) in the case of hard sphere interaction will suffice. For this purpose S(q) of a system of hard spheres may be obtained from the Percus-Yevick theory [66] generalized by Vrij and coworkers [67,68] to polydisperse systems. The solid line in Fig. 6 displays S(q) resulting for a system of hard spheres with a Gaussian size distribution characterized by a standard deviation of 7.5%. The main feature induced by polydispersity is the much weaker side minimum of S(q) as compared to the monodisperse case. Hence, a finite width of the size distribution will tend to smear out the oscillations of S(q) at higher q.

Very often SAXS-measurements need to be conducted at rather high concentrations to obtain a good counting statistics. The considerations related to S(q) now demonstrate that the SAXS-analysis can indeed be done at a rather high concentration provided the q-value is not too small.

To discuss this point in detail we again use the structure factor of a system of polydisperse hard spheres given by Vrij and coworkers [67,68]. Figure 7 displays the resulting scattering intensities normalzied to volume fraction calculated for different concentrations [51]. Here a Gaussian size distribution with a standard deviation of 10% has been assumed. For better comparison the scattering vector q has been scaled by R_g.

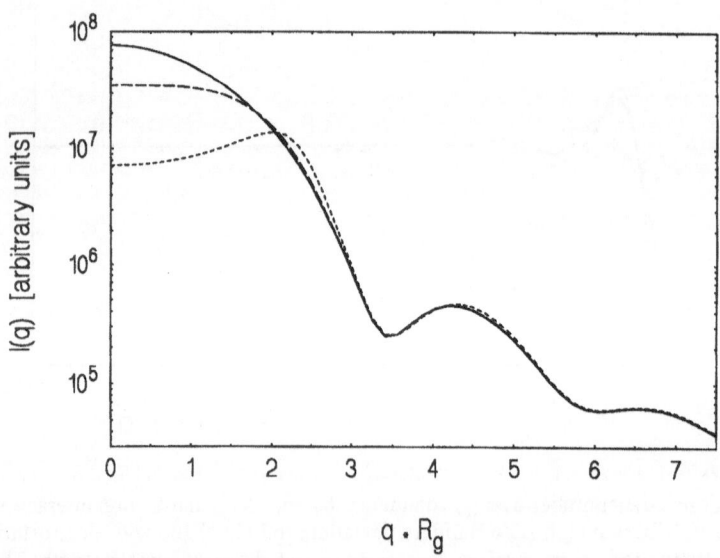

Fig. 7. Influence of particle interaction on SAXS-scattering curves: Calculated scattering intensities normalized to volume concentration for a system of hard spheres [51] of 80 nm diameter. The size distribution was assumed to be Gaussian with a standard deviation of 10%. Parameter of the different curves is the volume fraction of the spheres. *Solid line*: vanishing concentration (0%); *dashed line*: 10%; *dotted line*: 30%

It becomes obvious from this model calculation that the influence of $S(q)$ is restricted to the region of smallest angles. Even at a volume fraction of 30% the scattering curve superimposes with the result calculated for vanishing concentration when $q \cdot R_g > 3$. Therefore the SAS-measurements can be done at rather high concentrations if the region of smallest angles is of no interest. At these high concentrations the scattering intensities at high q can be determined with good accuracy, whereas low concentrations would lead to noisy data. In view of this fact, most of the measurements reviewed herein have been conducted using volume fractions between 5 and 10%.

2.3
Contrast in SAXS and SANS

The considerations of Sect. 2.1 have demonstrated the importance of contrast in a small-angle scattering experiment. Therefore it is interesting to compare the contrast of typical polymers achieved in a SAXS-experiment to the contrast in a SANS-experiment. Table 1 gathers the scattering length densities for some polymers used in emulsion polymerization together with the respective values of H_2O, D_2O, and 40wt.% aqueous sucrose solution. The scattering lengths were taken from Ref. [14] and the total coherent scattering length density can be calculated according to [14]

$$b_n = \sum_\alpha a_\alpha^{coh} / v \qquad (29)$$

where the summation extends over all atoms in volume v and a_a^{coh} denotes the coherent scattering length of the element α. For a meaningful comparison of contrast in SANS and SAXS experiments, explicit scattering length densities have been collected in Table 1 for the SAXS-case as well. For this purpose the electron densities normally used to characterize the contrast in SAXS-experiments have been multiplied by the Thomson factor r_0 as discussed above.

From Table 1 it is directly evident that the electron densities of the above polymers can be easily matched by a sucrose solution of proper weight concentration. In the case of poly(butadiene) the electron density is even lower than the electron density of water.

In the case of SANS the scattering length density of the medium can be changed by appropriate mixtures of H_2O and D_2O. The scattering length densities in Table 1 furthermore show that the difference in the scattering length density of particles and medium is considerably larger in the case of SANS which leads to a much stronger scattering intensity. SANS is therefore often indispensable when measurements at maximum contrast are required to determine $I_S(q)$ with highest accuracy. On the other hand, there is an incoherent contribution to the SANS-intensities [14] absent in SAXS. For this reason SAXS is often more suitable when looking at the region of high angles.

Table 1. Calculation of SAXS and SANS-scattering length densities for polymers

molecule	ρ [g/cm^3][1]	ρ_e [1/nm^3]	b_x (SAXS) [10^{10} cm^{-2}]	b_n (SANS) [10^{10} cm^{-2}]
H_2O	0.997	333.4	9.33	−0.56
40wt.% sucrose solution[2]	1.170	384.1	10.82	
D_2O	1.10	332.9	9.32	6.34
PS	1.05	339.7	9.51	1.41
PMMA	1.19	383.4	10.80	1.31
PBA	1.08	355.0	10.00	0.65
PB	0.97	323.7	9.12	0.45

ρ_e: electron density; the densities of the polymers necessary for the calculation of ρe have been
   taken from Ref. [84]
[1] taken from [84]
[2] taken from Ref. [47]
PB: poly(butadiene)

3
Experimental Techniques

A survey of different cameras used for small-angle analysis has been given by
Pedersen [69] which provides a practically complete overview of all systems
used up to now for SANS and SAXS. Here we shall focus on devices used for
SAXS-measurements. In particular, an extended discussion of the Kratky-camera [70, 71] will be given because this device allows very accurate SAXS-measurements using an ordinary X-ray generator. Most of the experimental investigations on polymeric latexes have been conducted using this type of SAXS-camera. Since this device works in slit collimation, the correction of the data will be
discussed in detail.

In addition, we will give a brief overview on other cameras which require X-
ray sources with high intensity thus necessitating the use of synchrotron radiation. It is evident that, up to now, such X-ray sources are not available on a routine basis. SAXS, on the other hand, will be shown to be a highly versatile tool
for the analysis of latexes and instruments allowing measurements by use of
conventional X-ray sources are therefore very useful.

Besides a description of instruments, the following section will also contain a
discussion of the treatment of data. This includes the removal of various contributions to the signal stemming from the suspension medium water and from the
density fluctuations of the solid polymer in the latex particles.

3.1
Slit Collimation

Since its introduction in 1954 by Kratky and coworkers [70,71] the so-called Kratky-camera has been used in the course of a great number of studies of the SAXS of polymeric and colloidal systems. Due to the ingenious design of the block-collimation system, the signal-to-background ratio can be optimized in this device. This gives one the opportunity to investigate weakly scattering samples, e.g. polymers in solution. The smearing of the data due to the slit collimation used by this device, however, is often thought to be a major disadvantage of this camera. It can be overcome by applying a variety of routines which have been developed in recent years [72]. These methods allow one to desmear the data without too much difficulty if the signal-to-noise ratio is sufficiently high. Also, in a number of cases the assumption of an infinitely long slit is justified and the slit-smeared data may be evaluated directly [1].

In the following, a description of an improved Kratky-camera [73] will be discussed together with an extended discussion of the treatment of data. This device is capable of measuring latex particles up to a diameter of 200 nm and reaches the q-range provided by SAXS-cameras which work in point collimation and use synchrotron radiation (cf. below [73]).

Design of the SAXS camera. Figure 8 shows a diagram of the SAXS-camera. As in the case of a conventional Kratky-camera [71] it consists of a block collimation system, a sample holder, and a primary beam stop. The intensity of the primary beam is measured by a moving slit device [75]. The intensity is recorded by a linear position-sensitive counter (Braun, OED-50m) the spatial resolution of which is given by approximately 80 μm.

The counter is attached directly to the camera avoiding an additional window. This change of the original design [71] is important because a window and a gap between the camera housing and the counter would cause considerable parasitic

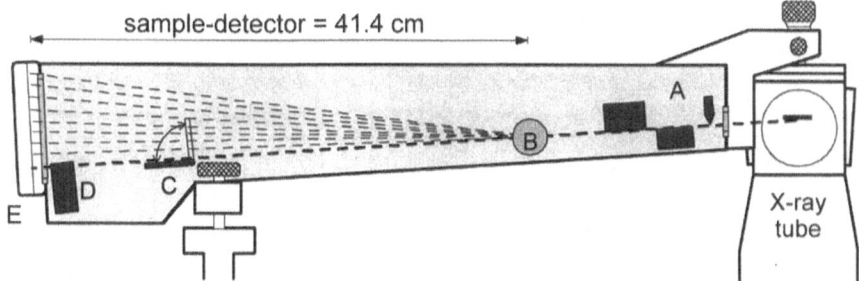

Fig. 8. Schematic drawing of the SAXS-camera (Kratky-design). *A*: block collimation system; *B*: capillary (sample holder); *C*: slit for measurement of intensity of primary beam; *D*: primary beam stop; *E*: one-dimensional counter

scattering. The camera is mounted on a conventional X-ray generator and Ni-filtered Cu K_α-radiation is used throughout all experiments.

The distance between the sample holder and the one-dimensional counter is 41.4 cm, which is approximately twice the value of the conventional design (23.6 cm) [70]. In principle, an enhancement of the distance sample-to-detector does not improve the resolution. For a given q-range the increased distance of the detector to the sample, however, leads to approximately twice the number of channels of the position-sensitive counter as compared to the conventional design. Hence, the region of smallest angles in which the measured intensity displays the strongest decrease may be measured and corrected much more accurately. As a consequence of this, the minima and maxima can be resolved clearly.

Besides these advantages the enhanced distance of the sample and the detector leads to an improvement of the angular resolution of the SAXS-device. This will become more evident when considering the principal sources of smearing for a slit-collimation system (cf. Fig. 9) [1,72]:

The finite dimensions of the primary beam lead to a smearing of the measured intensity $\tilde{I}(m)$ which may be expressed through the relation [1, 72]

$$\tilde{I}(m) = \int\limits_{-\infty}^{\infty} \int\limits_{-\infty}^{\infty} I_0(t,x) I[((m-x)^2 + t^2)^{1/2}] \, dx dt \qquad (30)$$

where $I_0(t,x)$ is the intensity profile of the primary beam (profile of the primary beam convoluted with the width of the detector in t-axis), and $I(m)$ is the scattering intensity which would result when measuring in point collimation. This

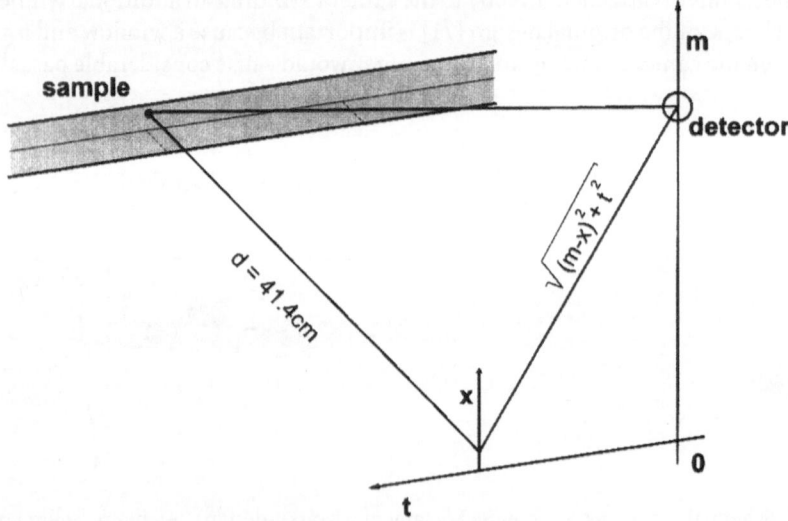

Fig. 9. Smearing of scattering intensity by slit-collimation: Definition of geometry and variables

profile may be decomposed into a part due to the slit length $P(t)$ and a part $Q(x)$ due to the width of the slit [76]:

$$I_o(t,x) = P(t)Q(x) \qquad (31)$$

Here $P(t)$ is the virtual profile in the t-direction in the plane of the sample which takes into account the finite divergence of the primary beam as well as the length of the detector in the plane of registration. The function $Q(x)$ gives the profile of the primary beam in the direction perpendicular to the slit length as measured in the plane of detection. For the block collimation system under consideration here, this profile is asymmetric and determines the smallest q-value accessible by the SAXS-camera. It results from the convolution of the intensity along the x-direction and the resolution function characterizing the finite resolution of the counter.

It is important to note that the latter effect strongly contributes to the smearing in the x-direction as expressed through the function $Q(x)$. Due to the geometry the primary beam has been spread out over twice the number of channels of the counter as compared to the conventional design. On the other hand, the distance between the channels of the detector remains the same. As a consequence of this, the finite spatial resolution of the position-sensitive detector has a significantly smaller influence on the scattering curve than is the case for the conventional design.

3.2
Treatment of Data, Desmearing

Data correction. Given the validity of Eq.(31) the smearing may be rendered by two one-dimensional integrals [76, 72]

$$\hat{I}(m) = \int_{-\infty}^{\infty} P(t)I[(m^2 + t^2)^{1/2}]dt \qquad (32)$$

and

$$\tilde{I}(m) = \int_{-\infty}^{\infty} Q(x)\hat{I}(m-x)dx \qquad (33)$$

Following Glatter and Zipper [76] the effect of $Q(x)$ must be corrected first. After the deconvolution of Eq.(33) which may be done conveniently by the algorithm of Beniaminy and Deutsch [77] the effect of slit length embodied in Eq.(32) can be corrected using the method of Strobl [78]. For the routines to be discussed here the raw data are fitted by an approximating spline function the coefficients of which are directly used for the deconvolution of integral Eq.(33) [77].

Because of the strong decrease of the scattering curves of latex systems which scales with q^{-4} at higher scattering angles as discussed above, the profile $P(t)$ can be approximated by a Gaussian which deviates only at large m from the experimental profile. Schelten and Hossfeld [79] have shown that the integral (32) may be solved analytically for $P(t)=Cexp[-a^2t^2]$. Thus, the correct intensity I(m) may be calculated from $\tilde{I}(m)$ and its first derivative [79]

$$I(m) = -\frac{1}{\pi C}\int_0^\infty \{\hat{\tilde{I}}'[(m^2 + t^2)^{1/2}] - 2a^2(m^2 + t^2)\hat{\tilde{I}}[(m^2 + t^2)^{1/2}]\}\frac{\exp[-a^2t^2]}{(m^2 + t^2)^{1/2}}dt \quad (34)$$

I(m) thus may be obtained from the smoothed experimental data through a numerical integration of Eq. (34). A critical comparison [73] of this method with the I(m) obtained from the experimental profile P(t) through use of the Strobl routine [78] shows that in the case of latexes Eq.(34) may be applied without problems.

A check of the above procedures was done by use of the iterative desmearing due to Lake [80]. Here the integral (30) is not solved directly but the full profile $I_0(t,x)$ is used to smear a trial function I(m). Comparison with the measured function $\tilde{I}(m)$ then leads to an improved trial function I(m) and the entire process is iterated until full agreement is reached. The advantage of this procedure is located in the fact that the first derivative of the measured curve is not necessary. Also, both the effect of slit length and slit width are taken into account at one step and the procedure of Lake [80] may therefore be used for an independent check of the deconvolution of integral (32) and (33). Thus, a critical comparison of three different methods of data correction could be done [73]. The results showed that the desmearing based on Eq.(34) can be used for the data deriving from latex particles.

Subtraction of solvent-induced background. Although the block-collimation system used in the camera discussed above suppresses most of the parasitic scattering at low angles, several other effects may lead to a considerable background which must be subtracted carefully from the measured intensities. To assess this problem in further detail, Fig. 10 gives a comparison of the different contributions to the measured scattering intensity of a polystyrene latex of 150 nm diameter [73].

The crosses in Fig. 10 display the SAXS-intensity due to the capillary filled with water as compared to the smeared intensity of the polystyrene latex [81] measured at a volume fraction of 18% (open circles) [73]. It is easy to see that at small scattering angles the intensity measured from the latex is partially due to the effect of the sample holder used for the measurement of the latex. The other important contribution is given by the intensity due to the density fluctuation of water (cf. below). Both contributions can be removed according to [82]:

$$\tilde{I}_L(m) = \tilde{I}(m) - (1-\phi)\tilde{I}_{bg}(m) - \phi\tilde{I}_{cap}(m) \quad (35)$$

where $\tilde{I}(m)$ denotes the measured intensity of the capillary filled with latex of volume fraction ϕ, $\tilde{I}_{bg}(m)$ is the intensity measured for a capillary filled with water, and $\tilde{I}_{cap}(m)$ is the respective quantity obtained from an empty capillary.

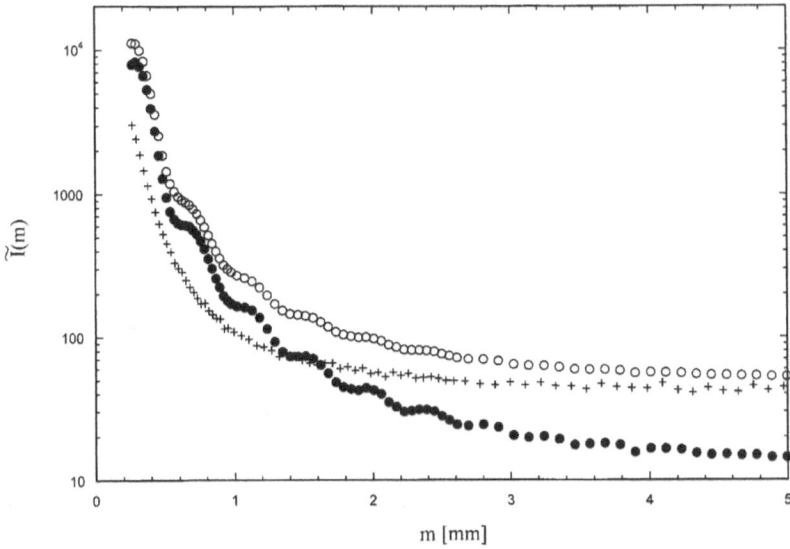

Fig. 10. Comparison of smeared scattering data obtained from a PS latex (volume fraction = 18%) used in Eq. (35). *Crosses*: Intensity $\tilde{I}_{bg}(m)$ of capillary filled with water; *empty circles*: intensity $\tilde{I}(m)$ of capillary filled with latex; *filled circles*: intensity $\tilde{I}_L(m)$ of latex

The filled circles in Fig. 10 display the resulting $\tilde{I}_L(m)$ which is considerably lower than $\tilde{I}_{bg}(m)$ at higher scattering angles. Another important feature seen in this comparison is the strong contribution of the background to $\tilde{I}(m)$ already in the region of the second side maximum. This is due to the small excess electron density of the polystyrene latex spheres in water [46]. The poor contrast of these particles necessitates a particularly careful subtraction of the background. It is obvious that the low contrast in the case of polystyrene spheres in water affects as well the measurement by a camera working in point collimation. In the case of PMMA which exhibits a much higher contrast in water [46, 47] much higher scattering intensities are obtained and the removal of the background due to the solvent provides no difficulty.

Interpolation of data. The application of the desmearing routines requires the interpolation, and if necessary the smoothing of the data by spline functions. Figure 11 shows that even for the case of the polystyrene particles having low contrast (cf. above) the fit by cubic splines amounts to an interpolation for the q-range in which the maxima of the scattering curve are seen. Only at highest scattering angles (see inset of Fig. 11) the scattering data exhibit a poor statistics and the fit of the spline functions is a smoothing of data indeed. It will become

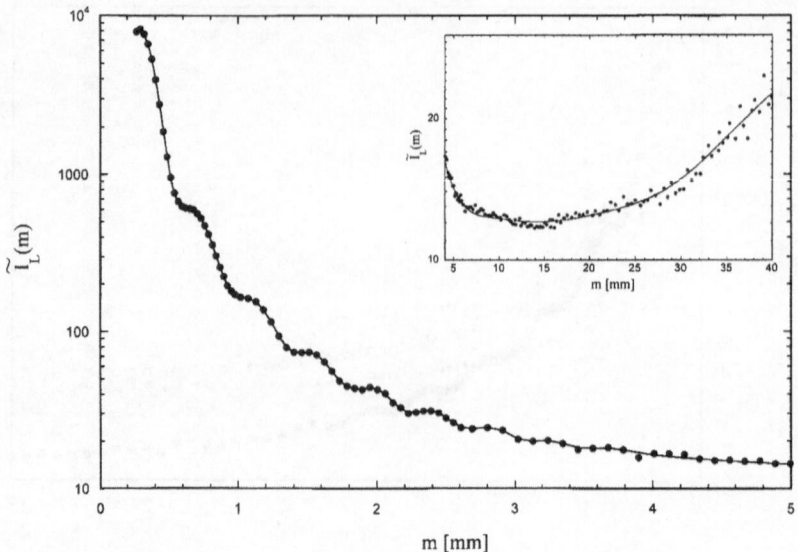

Fig. 11. Desmearing of data: Fit of $\tilde{I}_L(m)$ (see Fig. 10 and Eq. (35)) by cubic splines. The *inset* shows the smoothing of data at high q whereas the fit of the data at low q amounts to an interpolation

apparent further below that this angular region is of low importance for the structural investigation of the latex spheres.

Normalization of data; absolute intensities. The $\tilde{I}_L(m)$ data solely pertaining to the effect of the latex particles must now be normalized to the intensity of the primary beam. Here the moving slit method introduced by Kratky and Stabinger [75] has been used. The absolute intensity may then be derived taking into account the geometry of the camera as well as the absorption of the sample. The details of this procedure have been discussed repeatedly in literature (see e.g. Pollizi et al. [83]).

The main error incurred when determining absolute scattering intensities by the moving slit device is given by the insecure determination of the resolution function of the position-sensitive counter. Here a width of 80 μm has been used which is supplied by the manufacturer. To check this problem in more detail, the scattering intensities of water, toluene, and ethanol have been measured at 25 °C and compared to the theoretical result deriving from classical fluctuation theory [1]:

$$I(q=0) = k_B T \kappa_T \rho_e \tag{36}$$

where κ_T is the isothermal compressibility of the liquid, ρ_e its electron density and k_B and T have their usual meaning. The densities and the compressibilities have been taken from Ref.[84]. For all three liquids the absolute intensities calculated from (36) are higher by 6–8% which is within the limits of experimental error.

Desmearing of data. The normalized data obtained now may be subjected to the desmearing procedures as discussed in the preceding section. At first the deconvolution of Eq.(33) is effected through use of the procedure of Beniaminy and Deutsch [77]. Figure 12 displays the comparison of $\tilde{I}_L(m)$ (crosses) and intensity $\hat{I}_L(m)$ (empty circles) resulting after correction. At higher q the influence of the convolution Eq. (33) is practically negligible but the data show that this effect is of great importance in the region of the first three to four minima.

The subsequent correction for the slit length Eq.(32) leads to the scattering curve pertaining to point collimation (filled circles in Fig. 12). For a comparison of the corrected intensity thus obtained with models of the radial electron density the intensity due to the density fluctuations of the solid polymer must be subtracted. In principle, this contribution could be removed by the classical Porod plot [1,2] which assumes the scattering intensity to scale as q^{-4} and the background to be independent of q (see Sect. 2). In many cases the scattering curves exhibit oscillations up to high q-values, however (see e.g. Refs. [46–48]). In these cases the contribution due to density fluctuations should be determined from much higher q-values in which the form part of the scattering intensity may be

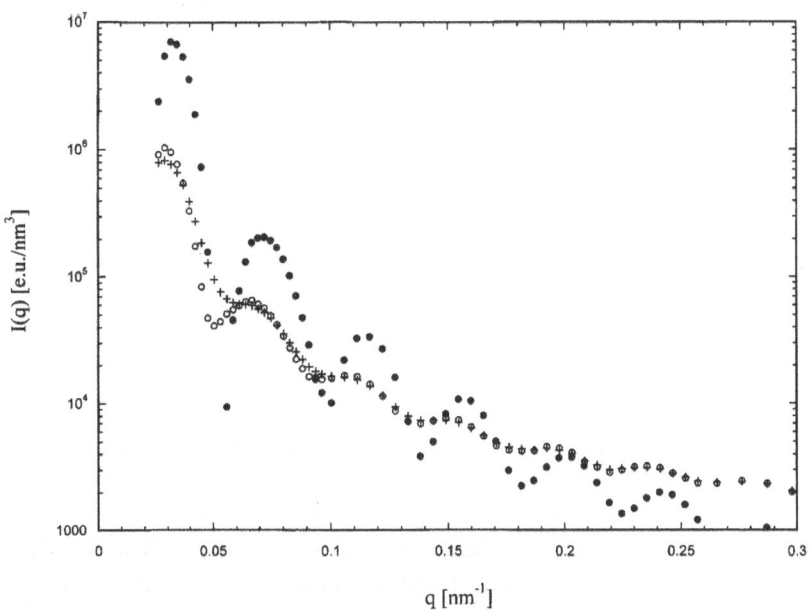

Fig. 12. Desmearing of $\tilde{I}_L(m)$ (see Figs. 10, 11 and Eq. (30); includes smearing by slit length and slit width): *Crosses:* $\tilde{I}_L(m)$; *empty circles:* $\hat{I}_L(m)$ (effect of slight width has been removed; cf. Eq.(33)), *filled circles:* desmeared intensities I(q)

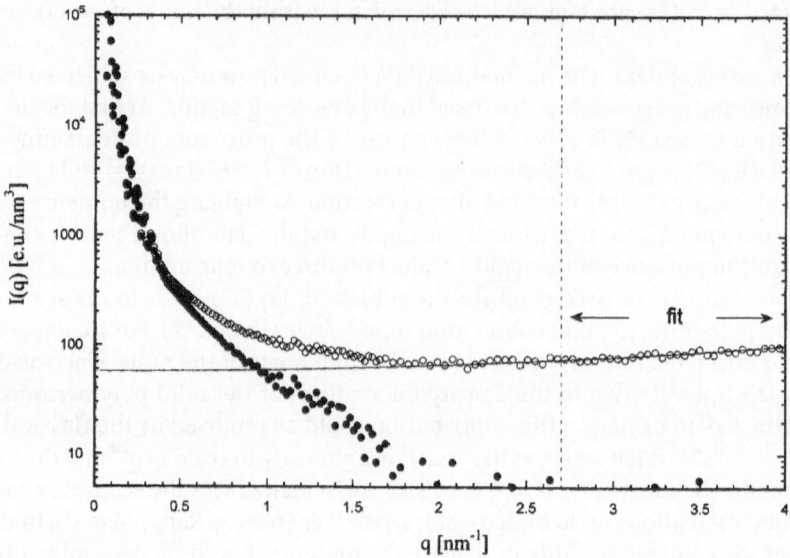

Fig. 13. Removal of the contribution I_{fluct} caused by density fluctuations of the solid polystyrene [85, 86]. *Empty circles*: Desmeared scattering intensity; the marked range shows the q-
range used for the fit of I_{fluct}; *filled circles*: scattering intensity after subtraction of I_{fluct}

safely dismissed. As discussed previously [47], the scattering curve at high q is
well described by the function $I(q)=A \cdot exp[Bq^2]$ (cf. Refs. [85, 86]). Fig. 13 shows
a fit to the region of high q together with the scattering intensity after subtraction of this part. It becomes obvious from this comparison that the fluctuation-
induced contribution becomes appreciable beyond $q=1$ nm^{-1} in the case of the
present system. A similar study devoted to micelles composed of blockcopoly-
mers [59] has come to the same conclusions. All investigations conducted so far
demonstrated that the scattering due to density fluctuations inside the particles
scales with the weight concentrations of the particles [59, 51]. This finding adds
further credibility to the above procedure of subtracting this part of the scattering intensity.

The scattering data thus corrected are solely due to the radial excess electron
density of the particles. Fig. 14 displays the measured intensity (filled circles) of
the polystyrene latex discussed in conjunction with Fig. 10. The solid line is the
fit of the experimental data by a core-shell model and a slightly asymmetric size
distribution ([46]; see below) taken from the analysis by ultracentrifugation
[87]. In terms of a Gaussian size distribution the polydispersity corresponds to
a standard deviation of 4.2%. The thin shell having a higher electron density
stems from the adsorbed surfactant used in the synthesis of the latex. This effect
and its detection by SAXS will be discussed further below (see Sect. 4.4).

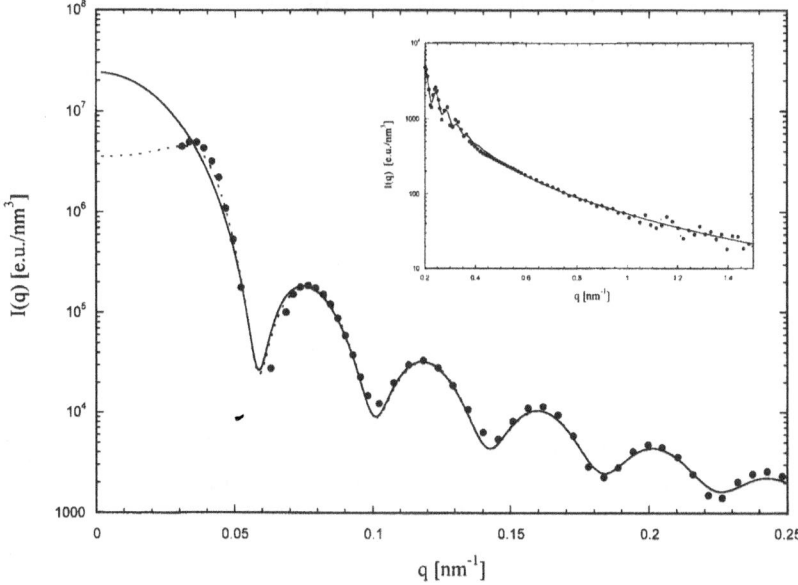

Fig. 14. Scattering intensity of a PS latex measured at a volume fraction of 18.6%. *Filled circles*: Measured intensity; *solid line*: calculated intensity $I_0(q)$ (cf. Eq. (1)); *dotted line*: intensity $I_0(q)S(q)$ calculated for a system of hard spheres (see Fig. 7)). The data have been taken from Ref. [73]

In this context the data shown in Fig. 14 demonstrate that the Kratky device described above is capable of measuring the SAXS-intensity function of a weakly scattering latex with a diameter of 150 nm.

3.3
Point Collimation

The foregoing section was devoted to the design of a camera working in slit collimation since such a device can be used together with a conventional X-ray generator. In principle, measurements conducted in point collimation would be preferable because here the problem of data correction as discussed in Sect. 3.2 are greatly alleviated. A review of the SAXS-cameras used up to now was given by Pedersen [69]. Recently, the high brilliance of X-ray radiation obtained by a synchrotron lead to the construction of SAXS-camera working in point collimation but having enough flux to allow measurements in very short time over a wide range of scattering angle. As an example of such an advanced device, the high-brilliance beam line at the European Synchrotron Radiation Facility in Grenoble recently introduced by Bösecke et al. [88] should be mentioned. Up till now, only a few measurements with this instrument as applied to latexes have been given in the literature [89]. It is expected that this camera will be applied more frequently in the near future. The restricted access to such a device, however, clearly points to the need of SAXS-cameras being mounted on ordinary X-ray sources.

3.4
Bonse-Hart-Camera; Ultrasmall Angle X-Ray Scattering (USAXS)

The design of all SAXS-cameras discussed so far is based on a system of properly chosen slits which defined the primary beam by geometric collimation. Such a collimation is restricted to scattering angles giving $q>0.01$ nm^{-1}; for smaller scattering angles the parasitic scattering of the defining slits becomes too strong and the signal-to-background ratio too poor to allow meaningful analysis of the scattered intensity. On the other hand, there is a gap in q-space between SAXS and light scattering around this q-value which is difficult to overcome by the design discussed in Sects. 3.2. and 3.3.

The key problem for reaching q-values below 0.01 nm^{-1} is to get an X-ray beam with a very small angular divergence without sacrificing its intensity. In 1966, Bonse and Hart [90] introduced a camera in which the primary beam is defined through a multiple reflection in a channel-cut crystal (see Ref. [69]). For perfect crystals each reflection leads to a further narrowing of the beam because its tails are drastically reduced while the intensity obeying the Bragg condition is preserved. This allows it to attain much smaller scattering angles (USAXS) and in consequence the study of considerably larger structures. The disadvantage of this device is the necessity of a strong X-ray source, preferably a synchrotron.

An extended discussion of the Bonse-Hart camera has been given by Chu and coworkers [91]; this reference as well as the review by Pedersen [69] also contain further citations on the subject. A further discussion of the technical problems has been given by Koga et al. [92].

The application of the Bonse-Hart camera to polymer latexes has been seldom up to now. The work of Chu et al. [91] already mentioned above has shown the great potential of USAXS for the structural study of latexes. Reus et al. [93, 94] and Ise at al. [95] have applied USAXS to the investigation of S(q) of latex particles in suspensions. Since USAXS-cameras are now available at synchrotron beam lines [96] more investigations pertaining to latexes and using this techniques are expected to appear in near future.

4
Structure of Latex Particles

4.1
Homogeneous Particles

First we shall discuss the SAXS-analysis of latex particles consisting of a homopolymer. This system serves as a check for the accuracy of the measurements and for a comparison with the general prediction of scattering theory as outlined in Sect. 2.

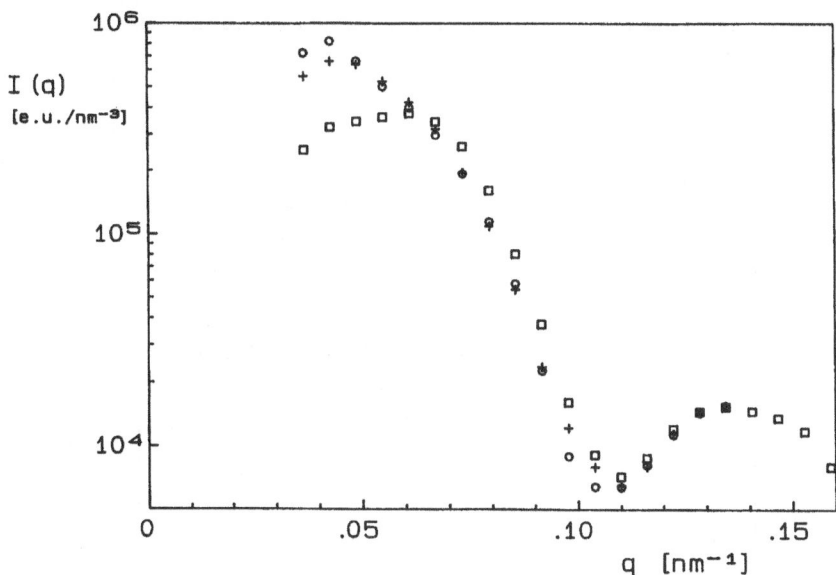

Fig. 15. Influence of $S(q)$ (see Eq. (1)) on the scattering intensity of a PS latex measured at different concentrations: squares: 18.7vol.%; *crosses*: 9.1vol.%; *circles*: 4.5vol.%. The data have been taken from Ref. [46]

As discussed there, the normalized scattering intensity is expected to become independent of concentration above a certain q-value related to the diameter of the particles. This can be shown easily by measuring the SAXS-intensities at different volume fractions. Figure 15 displays the normalized scattering intensities of a PS latex (diameter: 80 nm) measured at three different volume fractions [46].

From this comparison it can be seen immediately that interaction matters only at the highest volume fraction under consideration here (18.7%); for smaller concentrations the SAXS-curves practically merge beyond $q=0.06$ nm^{-1}. Even at a volume fraction of 18.7% the normalized intensities coincide for $q>0.11$ nm^{-1} in the case studied here. This is in full agreement with the theoretical deductions discussed in conjunction with Fig. 7.

Another important conclusion of basic theory is given by the fact that $I_0(0)$ scales with the square of the contrast $\bar{\rho}-\rho_m$ (cf. the discussion of Eq.(12)). As a consequence of this, the scattering curves of homogeneous spheres should be shifted parallel to the ordinate when changing the contrast. This is due to the fact that in this case the measured scattering intensity is solely given by the square of the form amplitude $B_0(q)$ (Eq.(14)).

Figure 16 shows that this behavior is observed indeed when measuring a PMMA latex at different contrast [52]. Here the data deriving from the highest contrast have been used for the fit; all other curves have been calculated from the respective contrast. Only at the lowest contrast and at higher scattering angles,

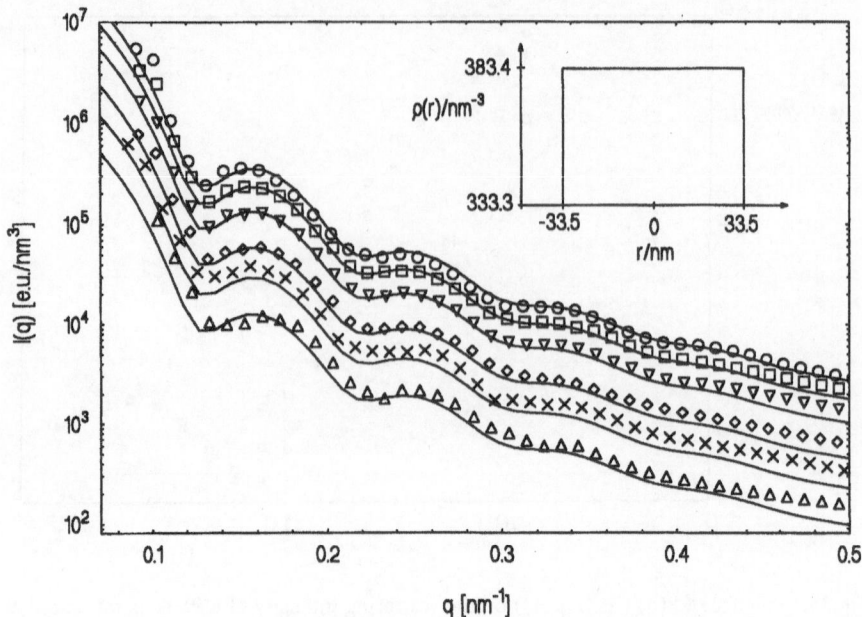

Fig. 16. SAXS-intensities of the unswollen PMMA-latex particles. The *solid lines* refer to the fit curves calculated assuming a homogeneous electron density within the particles (see *inset*). The numbers behind the symbols refer to the content (wt.%) of sucrose whereas the numbers in parentheses denote the contrast $\Delta\rho = \bar{\rho} - \rho_m/nm^{-3}$: O: 0% (50.1); □: 8% (40.7); ∇: 16% (30.8); ◊: 24% (20.9); Δ: 32% (9.5); x: 50% (-15.4). The data have been taken from Ref. [52]

the agreement is not quantitative anymore which gives a measure for the experimental error incurred in this region. Figure 16 also shows one of the problems of contrast variation: Due to the quadratic dependence of $I_S(q)$ on contrast, scattering intensities become quite low in the vicinity of the match point and therefore difficult to measure.

4.2
Core-Shell Particles

The potential of SAXS for a precise analysis of the radial structure of latexes can be discussed best when considering model particles consisting of a well-defined core and a closed shell of a second polymer. The particles analyzed by SAXS [45–49] have been prepared recently [45] by a seeded emulsion polymerization [97] of PMMA onto a polystyrene core having a narrow size distribution. The alteration effected by seeded emulsion polymerization can be seen directly in the analysis of the size distribution by ultracentrifugation [87], the resulting distributions are shown in Fig. 17. Besides the increase in radius when going from the

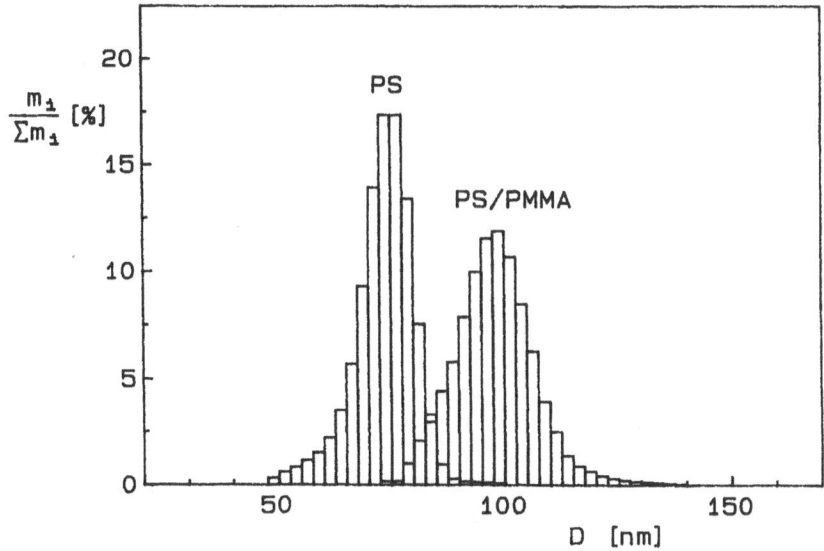

Fig. 17. Mass distribution of the diameter of the PS core latex and the PS/PMMA core-shell latex studied in Refs. [46] and [47] as determined by ultracentrifugation [87]. (Taken from Ref. [46])

PS-core particles to the core-shell latices, Fig. 17 also shows the slightly asymmetric size distribution found for rather narrowly distributed latex systems. For a highly precise analysis of the SAXS-data the directly measured distributions should be used instead of the Gaussian distribution often used in the analysis of colloidal particles [47].

Figure 18a,b gives the key result showing the scattering intensities of this latex at different contrasts taken from Ref.[47]. Figure 18a displays the full set of data in which each curve has been shifted vertically by a factor of ten for the sake of clarity. Figure 18b gives the unshifted intensities for three different contrasts.

The pronounced alterations effected through change of the electron density of the medium are evident. In particular, the isoscattering point indicated in Fig. 18b by an arrow leads to an outer radius of 91 nm in good agreement with the value obtained by ultracentrifugation (92.3 nm).

The analysis [47, 49] of the scattering intensities reveals a well-defined concentric core-shell particle. The data taken near the match point (filled circles in Fig. 18b) show furthermore that there is a finite contribution to the SAXS-intensity at q=0 even near vanishing contrast. This points directly to a polydispersity of the average contrast, mainly caused by the variation of the thickness of the shell [47]. Another important point of this analysis is the interface between the core and the shell. Here the interfacial region between the two immiscible polymers was found to be very small (<4 nm [49]).

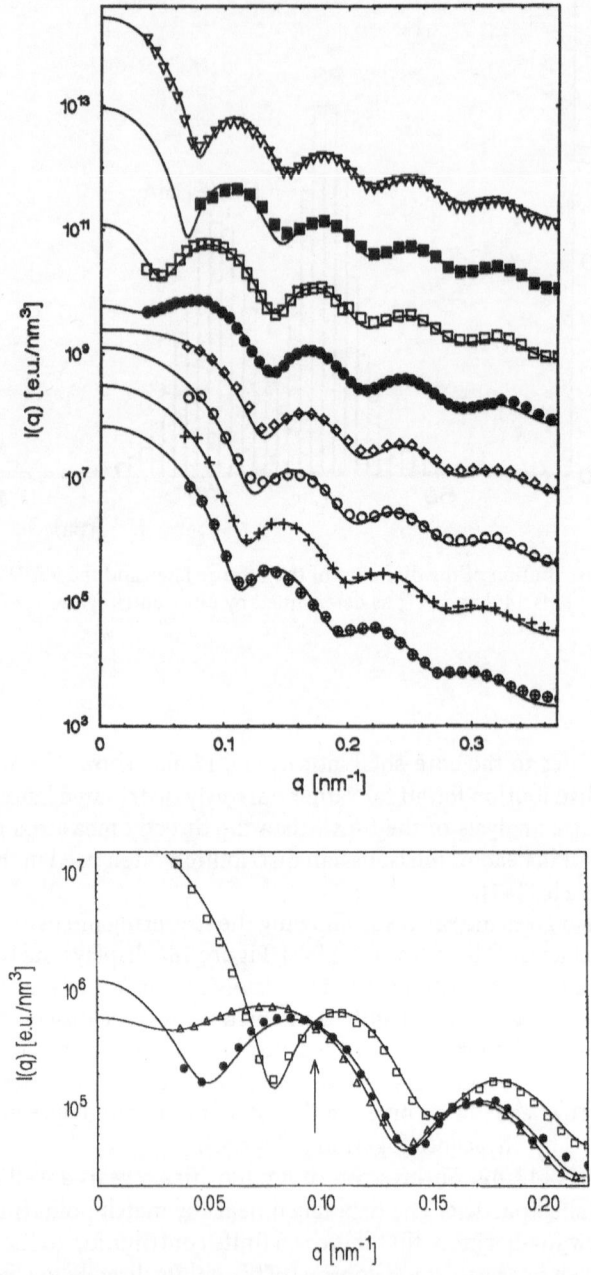

Fig. 18. SAXS-intensities for a core-shell latex at different contrast (taken from Ref. [47]). **a** q<0.4 nm^{-1}; for the sake of clarity the subsequent curves have been shifted by a factor of 10, the values at the ordinate belong to measurements at 50 wt.% sucrose. **b** unshifted curves at q<0.25 nm^{-1}. The contrasts of the different measurements are (in electrons/nm^3): ∇: 24.5; ■: 14.3, ❑ : 4.0; ●: 0.; ◊: -6.3, O:-16.5, +: -26.8, ⊕: -39.6

The fit shown in Fig. 18 is restricted by a number of experimental parameters such as electron density of the polymers, and the concentration of the particles. It must be noted that absolute intensities have been used here. Hence, the number density of the particles is fixed and cannot be used as a fit parameter. These constraints lead to the elucidation of the radial structure of the particles with a resolution of a few nm.

It must be noted that the process of seeded emulsion polymerization does not lead to an equilibrium structure. Hence, the sharp interface between PS and PMMA observed in the above core-shell particles cannot be explained by thermodynamic arguments. A possible mechanism may be sought in the adsorption of oligo(methylmethacrylate) radicals from the water phase onto the PS-seed particles [45]. The temperature of the seeded emulsion polymerization (80 °C; [45]) is well below the glass transition temperature of polystyrene and the adsorbed chains bear a sulfate endgroup. The adsorbed oligomers will therefore remain at the surface of the core particles and in consequence there is no extended interface between PS and PMMA in these .particles.

An extension of these investigations to a latex synthesized by the absorption method, i.e., by a seeded emulsion polymerization where the core latex was swollen with the second monomer methylmethacrylate prior to polymerization, showed a gradual transition from the PS core to the PMMA shell [48, 49]. Figure 19 shows the resulting scattering curve at highest possible contrast (in water) whereas Figure 20 displays the dependence of the scattering intensity on contrast.

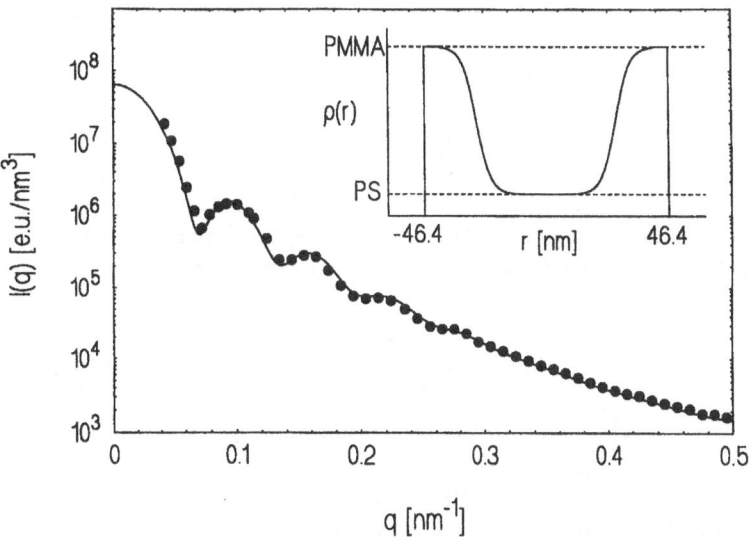

Fig. 19. SAXS intensity measured from a core-shell system with diffuse interface. The *circles* denote the experimental result whereas the *solid line* give the intensity calculated with the radial electron density profile shown in the inset. The data have been taken from Ref. [49]

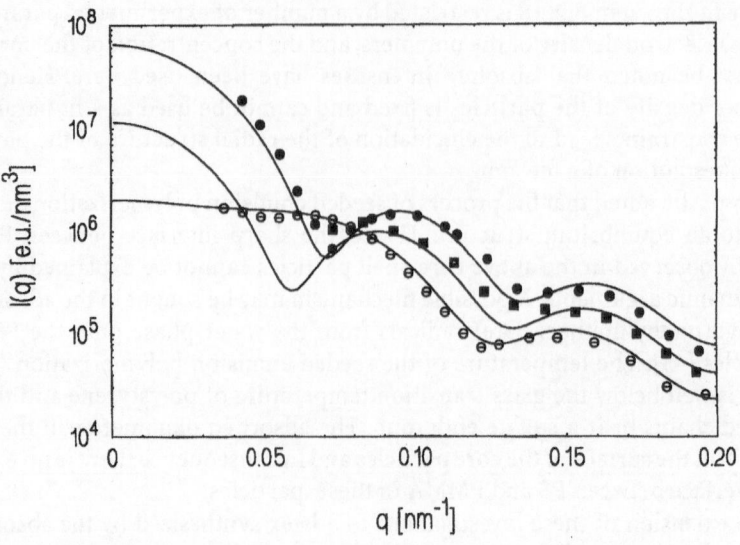

Fig. 20. SAXS intensity measured from a core-shell system with diffuse interface. The *symbols* denote the experimental result whereas the *solid line* give the intensity calculated with the electron density shown in the inset in Fig. 19. The contrasts of the different measurements are (in electrons/nm³): ●: 36.4; ■: 15.9, q: θ. The data have been taken from Ref. [49]

It is obvious from Fig. 20 that the crossing point is less pronounced than in the case of the core-shell latex discussed above (cf. Fig. 18b). The analysis of the internal interface [49] points to a morphology where the PS-phase is separated from the PMMA-phase by an irregularly shaped interface.

An analysis of a latex consisting of a PS core and a polytribromostyrene shell has been presented recently by Bianco et al. [98]. From SAXS-measurements at a given contrast these authors concluded that the particles under consideration exhibit a defined core-shell structure. By using brominated polymers the contrast could be enhanced considerably which facilitated the analysis to a certain extend.

In none of the studies discussed above, could the radius of gyration be obtained. The reasons for this are solely located in the restricted q-range of ordinary SAXS-cameras. On the other hand, the study of this quantity obtained from composite particles would certainly be highly useful (cf. Sect. 2.1). It appears that such an investigation would merit the application of USAXS-cameras (cf. Sect. 3.4) which would provide data at the respective low q-values.

In general, the above discussion has shown that the interface between two immiscible polymers is certainly an interesting point which needs further investigation. It is evident that SAXS is a highly useful tool for this purpose because this method can directly be applied to the latex without any prior staining or choice of special monomers, as e.g., deuterated monomers.

4.3
Structure of Swollen Latex Particles

The foregoing discussion has clearly revealed that the structure of latex particles deriving from emulsion polymerization depends strongly on the mode in which the second monomer is introduced. A question intimately connected to seeded emulsion polymerization is related to possible inhomogeneities in swollen latex particles due to the "wall-repulsion effect" [99, 100]. Emulsion polymerization can be used to obtain high molecular weight polymers whose radii of gyration may be of the order of the radius of the latex particles. If such a particle is swollen by a solvent, the polymer chains tend to avoid the surface of the particle because of their restricted conformations near a wall. Therefore the polymer concentration near an impenetrable surface is diminished.

The wall-repulsion effect thus effected by purely entropic reasons belongs to the classical problems of polymer physics. Early expositions of the theory of the wall repulsion effect are due to Cassassa [99,100] and to de Gennes [101, 102]. More recently there has been a considerable number of theoretical studies of the problem [30, 103–107]. The theories based on the statistics of single chains near hard walls must be refined when considering the wall-repulsion effect for the high polymer concentration: the high osmotic pressure of the concentrated polymer solution tends to push the polymer segments towards the wall and in consequence diminish greatly the wall-repulsion effect [106].

Moreover, Evers, Ley and Hädicke [108, 109] showed that enthalpic effects may modify the local structure of swollen particles as well and must be taken into account for a quantitative analysis of the problem. The segmental density of a polymer solution near an interface to the water phase thus will depend on a subtle interplay of entropic and enthalpic forces.

In a latex particle, the wall-repulsion effect would lead to a core-shell structure where the solvent would be enriched in an outer shell. This requires, however, that there is a strong enthalpic penalty for the polymer chains when entering the water phase. If, on the other hand, the swelling agent is much less polar than the polymer, its concentration near the interphase to water may be depleted and a polymer-rich layer will be formed due to enthalpic reasons.

Another counter-argument against a strong wall-repulsion effect derives from the so-called surface anchoring effect [110]: water-soluble initiators used in emulsion polymerization will lead to polar endgroups of the polymer chains. These endgroups are expected to be affixed near the surface of the particles and counteract the entropic repulsion.

Up to now, the experimental studies of the wall-repulsion effect conducted on polymer latexes came to conflicting statements. On the one hand, Linné et al. [24], Yang et al. [20, 21] as well as Dabdub et al. [111] concluded from their SANS-experiments that there is a marked effect for a certain ratio of the radius of gyration of the polymer chains to the particle diameter. On the other hand, Mills et al. [23] concluded from a set of SANS-experiments that the wall-repulsion effect is negligible. These authors used contrast variation to elucidate the radial structure of swollen latex particles in order to detect the enrichment of the swelling agent near the interface.

As shown recently [52, 55], SAXS together with contrast variation is the method of choice for a detailed study of this problem. The electron densities of low-molecular-weight solvents as e.g. methyl methacrylate (MMA) are much smaller than that of solid PMMA. It should thus be possible to detect even a small layer at the surface of the particles in which the solvent is enriched. Therefore the wall-repulsion effect should lead to a thin shell near the surface of the particles in which the polymer concentration is depleted.

A different situation arises when studying PMMA-latexes swollen by a non-polar monomer like styrene which exhibits at ambient temperature a much lower solubility in water (0.2 g/l) than MMA (15.9 g/l) [55]. Styrene has a very low electron density (see Table 1) in comparison to solid PMMA and both an enrichment or a depletion of this monomer in the surface layer are easily discernible in a SAXS-experiment [55]. In comparison to the above system PMMA/MMA a critical test of the influence of entropic versus enthalpic forces becomes possible: if the entropic wall-repulsion effect prevails styrene should be enriched in a surface layer. Because of the lower electron density of styrene this surface layer must exhibit a lower electron density than the core of the particle. If, on the other hand, the unfavorable enthalpic interactions between styrene and water are decisive, the more polar polymeric component PMMA should be enriched in a surface layer. In that case a surface layer with an enhanced electron density is expected.

All theoretical deductions discussed so far show that a surface layer caused by entropic as well as by enthalpic forces must be very thin and can only be of the order of a few nanometers. From the discussion in Sect. 2 it became evident that the consequences of this small alteration of the radial density profile for the scattering curves show up only in the immediate vicinity of the match point. At high contrast the curves are shifted parallel to the ordinate as is the case for homogeneous particles and SAXS-experiments conducted under these conditions would be misleading. Near the match point, however, the maxima are shifted in a characteristic fashion parallel to the abscissa which allows to distinguish between a homogeneous particle and a particle with an inhomogeneous density distribution. This point has been discussed already in Sect. 2 (cf. also the discussion in Ref. [52]).

In the course of the SAXS-studies of a PMMA latex swollen with MMA [52] or the unpolar styrene [55] it became obvious that both a depletion as well as an enrichment of the polymer in the surface layer may be observed: Fig. 21 displays the SAXS-scattering intensities obtained from a PMMA latex swollen by MMA [52].

At high contrast there is only a parallel shift of the scattering curves (triangles and empty circles in Fig. 21). This would point to a homogeneous radial electron density. Measurements in the vicinity of the match point, however, show that the swollen particles cannot be homogeneous: at positive contrast the extrema of the curves are shifted away from the ordinate whereas beyond the match point

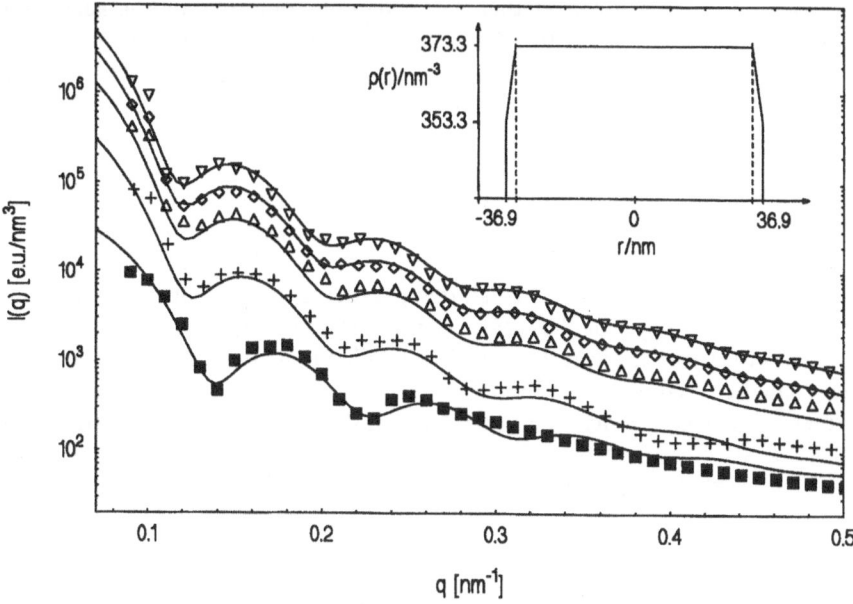

Fig. 21. SAXS-intensities of PMMA-latex particles swollen with MMA (Ref. [52]) measured at different contrasts. The *solid lines* refer to the fit curves calculated assuming the radial electron density shown in the inset. The numbers behind the symbols refer to the content (wt.%) of sucrose whereas the numbers in parenthese denote the contrast $\Delta\rho = \bar{\rho} - \rho_m /e^-\cdot nm^{-3}$: ∇: 0% (38.3); $\Diamond$: 8% (28.9); $\triangle$: 16% (19.0); +: 24% (8.7); ■:28% (2.0)

there is a pronounced shift towards the ordinate again. Note the low intensities around the match point which clearly present an experimental difficulty and require low parasitic scattering. The result points to a small shell ($d\approx 2$ nm) in which the swelling agent is enriched. In this case the wall-repulsion effect seems to prevail but a quantitative comparison with theory seems to be difficult in view of the smallness of the surface layer.

Doubts may be raised in this context whether the addition of sucrose to the latex may have influenced the above result. One might speculate that the sucrose may penetrate into the outer regions of the latex particles which have been plastified by the swelling agent. This argument can be easily refuted when considering the much higher electron density of sucrose as compared to the swelling agents or PMMA. A thin layer in which the sucrose solution would enter must have a much higher electron density whereas the SAXS-results shown in Fig. 21 must be explained by a layer of lower electron density. Furthermore, in this case the absolute electron density of the layer should rise with increasing sucrose concentration in the medium which is not observed. In contrast to this, all results obtained at widely different contrasts can be explained by a single model of the radial electron density. These considerations, together with additional experiments to be discussed further below, demonstrate that adding sucrose or glycerol, in order to change the electron density, may be used when studying latex systems. Any interaction of these agents with the particles would become evident when evaluating the data obtained at different contrasts (cf. also the discussion of this point in Ref. [59]).

In this context the question arises whether the results obtained in Ref. [52] are contradicting the work of Mills et al. [23] who did not find a depleted layer in their SANS-experiment of PS latex particle swollen by toluene. In accord with an earlier investigation along these lines by Goodwin et al. [16] these workers observed the scattering curves to be shifted parallel to the ordinate when lowering the contrast (cf. Fig. 5 of Ref. [23]). The lowest contrast, however, is still quite high and therefore these authors only concluded that significant inhomogeneities could be ruled out. This is in full agreement with the theoretical deductions presented in Sect. 2 which demonstrated that the detection of small inhomoge-

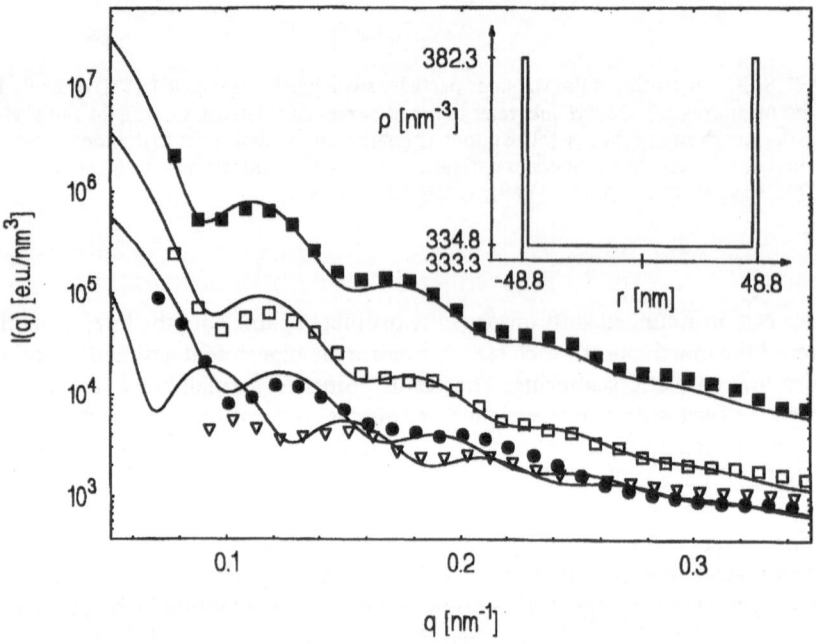

Fig. 22. Contrast variation measurements from the PMMA latex swollen with styrene at a volume ratio PMMA : styrene of 42 : 58. The curves refer to the following concentrations by weight of sucrose in the dispersion medium, whereas the number in parentheses denote the average contrast in nm^{-3}: (∇) 0% (4.4), ($\bullet$) 8.0% (-5.2), ($\square$) 16.0% (-15.7), ($\blacksquare$) 40.0% (-47.1). The *solid lines* refer to the fit curves calculated by assuming a radial electron density distribution within the particles as shown in the inset (water taken as a reference). Core radius: 47.8 nm; shell thickness 1.0 nm, volume average electron density 337.7nm^{-3}. The data have been taken from Ref. [55]

neities demand a close approach to the match point. At higher contrast the small contribution of $\varepsilon(q)$ (cf. Eq.(15)) is masked by the form part $B_0(q)$ (Eq.(14)).

As already mentioned above, the crucial test of the theory of Evers et al. [108, 109] may be performed using the system PMMA/styrene. Here a SAXS-study conducted along the lines discussed above showed indeed that here the enthalpic effects prevail. The key result is displayed in Fig. 22 giving the SAXS-intensities of PMMA particles swollen by styrene and measured at four different contrasts [55]. The analysis demonstrates that in this case a thin surface layer having a higher electron density must be assumed when modeling the SAXS-data (cf. inset of Fig. 22). Here obviously the nonpolar styrene tries to avoid the surface to the water phase and the concentration of PMMA which has a much higher electron density than styrene is enriched in the surface layer. It is thus obvious that the free enthalpy must be regarded when discussing the radial structure of swollen latex particles; considerations solely based on entropy may lead to erroneous conclusions.

4.4
Surface of Latex Particles; Adsorption of Surfactants

The general discussion of contrast variation in Sect. 2 has demonstrated that the investigation of the radial structure of latex particles often requires low contrast. The studies of swollen latexes discussed previously provided a good example for this assertion. For many systems this requires the addition of agents like sucrose or glycerol to the serum. In the case of PS-latexes, however, the electron density of this material is nearly matched by the electron density of water ([46]; cf. Table 1). These particles are thus near the match point and no agents need to be added to the serum. Ionic surfactants as e.g. sodium dodecylsulfate (SDS) exhibit a much higher electron density due to their polar head group. The same holds true for non-ionic surfactants like Triton X-405 (Scheme 1) in which the poly(ethylene oxide) chain makes the main contribution to electron density.

Thus, adsorption of small amounts of these surfactants can be monitored conveniently by SAXS because in the case of PS particles the main scattering intensity arises from the surface layer. In consequence, the scattering intensity of PS-particles covered by surfactant molecules strongly increases compared to the uncovered particles. Therefore PS latexes present ideal model systems for studying the adsorption of surfactants and polymers on colloidal particles from solution [53].

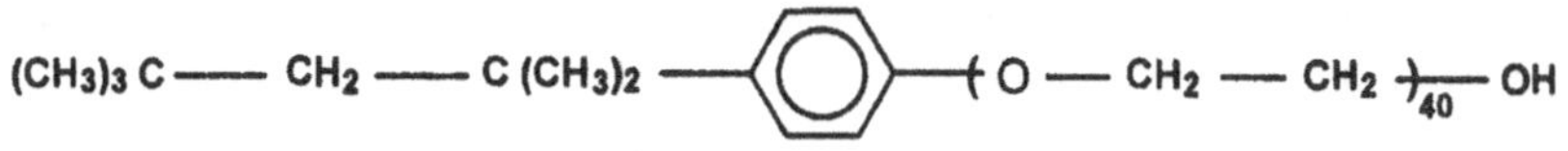

$$(CH_3)_3C \text{---} CH_2 \text{---} C(CH_3)_2 \text{---} \bigcirc \text{---} (O \text{---} CH_2 \text{---} CH_2)_{40} \text{---} OH$$

Scheme 1

Here again, it is interesting to compare SAXS to SANS as applied to the investigation of adsorbed surfactants. Studies of adsorption on colloidal particles have been conducted previously by Ottewill and coworkers employing SANS [27–29]. There the scattering of the particles could be matched by appropriate mixtures of H_2O and D_2O. This allows the application to inorganic colloids which cannot be matched with regard to their electron density. SANS has also the advantage of being capable to reach smaller q-values to obtain the radius of gyration. SAXS, on the other hand, can be extended to high q-values without being hampered by an incoherent contribution to the scattering intensity. A combination of both methods therefore seems to be the best way to fully elucidate the adsorption in a wide-spread variety of systems (cf. also Sect. 2.3).

Figure 23 shows the alterations of the SAXS-scattering intensity of a PS-latex when Triton X-405 is added [53]. Even small amounts of added surfactant cause a pronounced shift of the side maxima towards smaller scattering angles. This shift continues until saturation of the latex surface is reached at approximately 115 mg per gram PS. When more surfactant is added (cf. Fig. 23b) the side maxima remain at their position which they attained at the point of saturation. A strong increase of the scattering intensity at higher scattering angle ($q>0.3$ nm^{-1}) appears, however (see below).

Following the arguments expounded in Sect. 2, the shift of the side maxima are shown to be caused by the gradual build-up of a pronounced core-shell structure due to the adsorbed surfactant. An additional proof can be obtained by investigating the covered particles by contrast variation. To be sure that the added agent has a negligible influence on the process of adsorption, two different agents, namely sucrose and glycerol have been used to raise the electron density of the medium [53]. Here a precise comparison of the SAXS-intensities obtained at the same contrast which had been adjusted either by glycerol or by sucrose showed that the gross features of the covered particles turn out to be the same (see the discussion of Fig. 7 in Ref. [53]). Any profound alteration induced by one of these agents would show up immediately when making this comparison. This result adds further credibility to the deductions obtained so far by contrast variation and demonstrates that neither glycerol nor sucrose introduce any disturbance to the systems under consideration here.

It has been demontrated in Ref. [53] that beyond the point of saturation of the surface of the latex particles there is the formation of free Triton micelles. This is obvious when looking at the region of higher scattering angles (cf. Fig. 24). Adding more and more surfactant is followed by an increase of the scattering intensity in this region. Since the micelles are much smaller than the latex particles their scattering signal extends to higher q-values which can be easily separated from the signal of the covered particles.

In principle, the scattering intensity of a binary mixture of latex particles and micelles consists of three terms, i.e., there is a cross term in addition to the scattering intensities of the single components [1, 5]. The theoretical considerations propounded in Sect. 2, on the other hand, assumed mainly systems consisting of

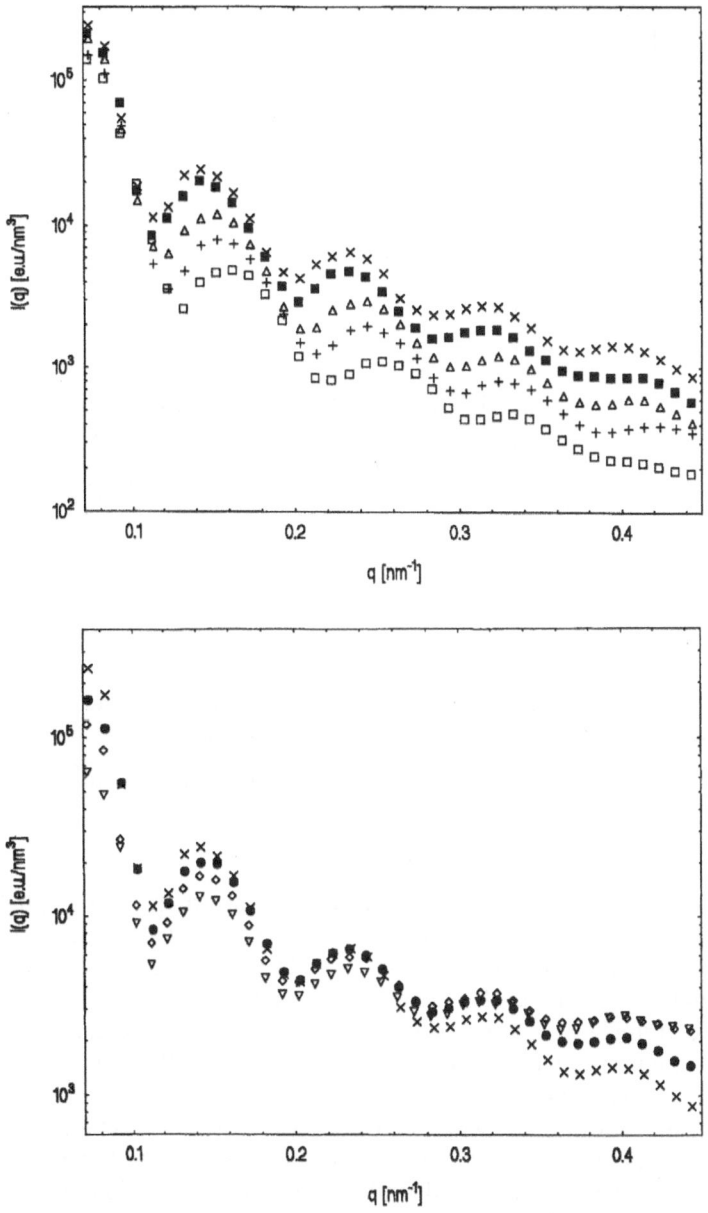

Fig. 23. a Scattering curves of the PS latex measured in water after addition of different amounts of Triton X-405. The curves refer to the following Triton concentrations expressed by mg Triton per g PS: □: 0 mg/g; +: 28 mg/g; Δ: 61 mg/g; ■: 115 mg/g (saturation); X: 195 mg/g. **b** Scattering curves of the PS latex measured in water after addition of different amounts of Triton X-405 beyond the saturation of the particle surface. The curves refer to the following Triton concentrations expressed by mg Triton per g PS: X: 195 mg/g; ●: 364 mg/g; ◊: 694 mg/g; ∇: 1020 mg/g. All data have been taken from Ref. [53]

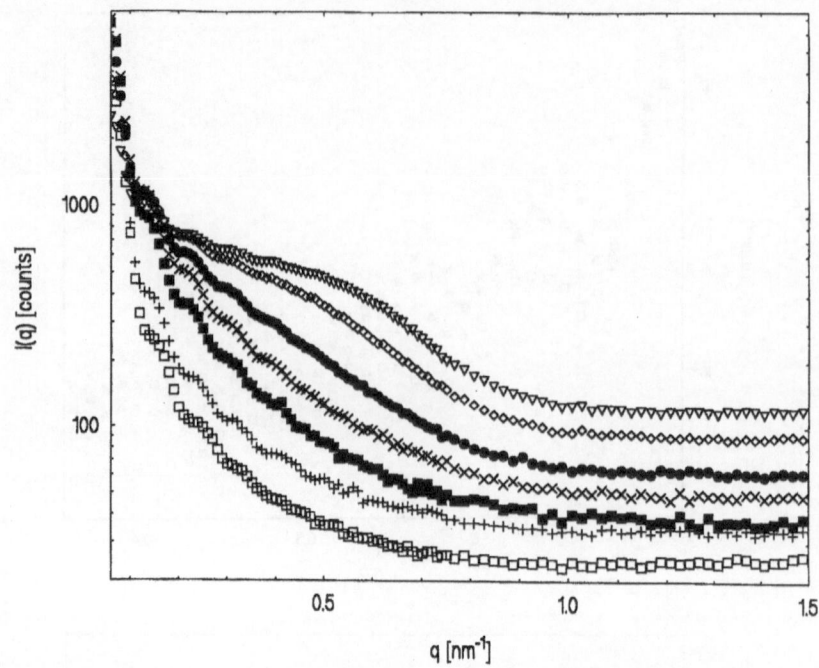

Fig. 24. Smeared scattering curves (range of high q-values) of the PS-latex after addition of different amounts of Triton X-405 measured in water. The curves refer to the following Triton concentrations expressed by mg Triton per g PS: □: 0 mg/g; +: 28 mg/g; ■: 115 mg/g (saturation); X: 195 mg/g; ●: 364 mg/g; ◊: 694 mg/g; ∇: 1020 mg/g. All data have been taken from Ref. [53]

one species. Here the problem is alleviated by the fact that the latex spheres have much greater diameters than the surfactant micelles. Furthermore, the cross term which vanishes in an ideal solution will be small for the rather dilute systems under consideration here. Hence, the scattering intensity at small angles will be governed by the large latex particles whereas in the region of higher angles the intensity is determined by the signal of the small component.

Based on these arguments the intensities in the wide angle region can be compared quantitatively with the respective signal of the free micelles. It turned out that the excess of the surfactant forms free micelles not attached to the surface of the covered particles [53]. If the micelles or a part thereof would stick to the surface, the scattering intensities at wide scattering angles would be weakened considerably.

The results of the investigation of reference [53] may be summarized by concluding that i) Triton X-405 is adsorbed up to a well-defined saturation point in a monomolecular layer, ii) the layer of adsorbed surfactant is rather dense (1–3 nm) which requires a part of the polar moieties to be located near the surface, and iii) the surfactant forms free micelles beyond the point of saturation.

These investigations can be extended to include the problem of competitive adsorption of two different surfactants, namely Triton X-405 and sodium dodecylsulfate [54]. Such competitive adsorption of a non-ionic and an ionic surfactant has been studied earlier by Kayes [112] and by Kronberg and coworkers [113, 114]. SAXS is suitable to investigate the mutual replacement of one surfactant adsorbed on the surface of PS-particles by the other one. This is made possible by the fact that SAXS can distinguish between the free micelles of both surfactants: Fig. 25 shows the SAXS-intensities obtained from solutions of both surfactants [54]. It can be seen that the Triton micelles exhibit the maximum scattering intensity at small angles whereas the SDS-micelles display a maximum at large angles. The peculiar scattering pattern of the SDS-micelles has been explained by Zemb and Charpin [115]. The high electron density of the polar head groups is counterbalanced by the low electron density of the hydrocarbon chains constituting the core of the micelles. Therefore the mean electron density of these micelles are nearly at the match point (see Sect. 2).

For a mixture of both surfactants the scattering curves indicate the formation of mixed micelles. This becomes obvious when studying the respective mixtures of both surfactants in water by SAXS [54]. The smeared SAXS-intensities of mixed micelles may easily be distinguished from the respective results obtained from the pure systems (cf. Ref. [54]).

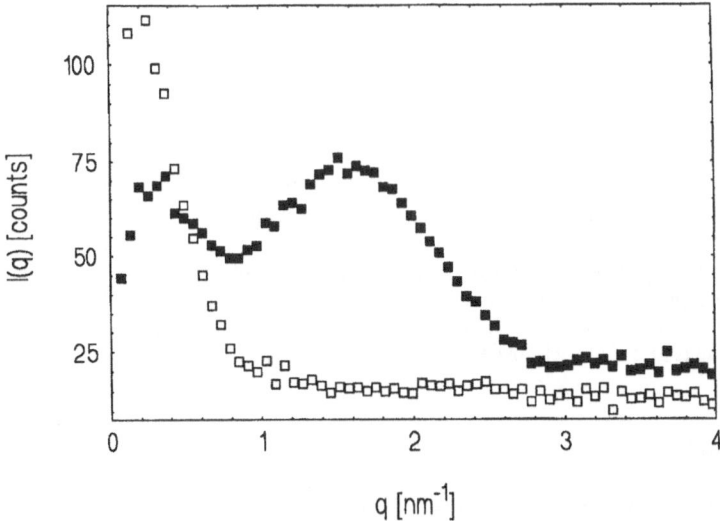

Fig. 25. Monitoring competitive adsorption of two surfactants on the surface of PS latex particles [54]. Scattering intensities of the free Triton X-405 micelles (□) and of the free SDS micelles (■) (concentration:1 vol.%). Both signals can be easily distinguished which allows to scrutinize the competitive adsorption of both surfactants on a PS latex by SAXS

As the consequence of these propitious facts, the replacement of the previously adsorbed SDS by Triton X-405 can be monitored by the build-up of the characteristic scattering signal of the micelles consisting of SDS when Triton is added. The reverse process can be monitored as well: Adding an excess of SDS to a PS-latex covered fully by Triton X-405 leads to characteristic changes of the scattering curve in the small-angle region as well as at high scattering angles. Again a similar analysis as outlined above leads to the conclusion that a rather large excess of SDS is able to replace the Triton X-405 from the surface [54].

These investigations demonstrated that Triton X-405 is much stronger bound to the latex surface of a polystyrene latex than SDS as expected. Moreover, competitive adsorption of these surfactants obviously is an equilibrium process because the scattering curve of a latex covered with the surfactants can be shown to be invariant against the mode of preparation. For a fixed composition of Triton and SDS it does not matter which surfactant has been added first.

5
Conclusion

It has demonstrated that SAXS is a highly useful method for the study of polymer latexes. When monitoring the scattering intensities in the vicinity of the match point, SAXS is capable of detecting minute variations of the radial electron density as induced by e.g. swelling of the particles by a solvent or by adsorption of surfactants. The contrast between most of the polymers used in emulsion polymerization is sufficient. The overall contrast of typical composite particles as e.g. core-shell latexes is still low enough to be matched by adding sucrose to the serum. Therefore no special monomers marked by deuteration are needed. The latex can directly be analyzed without prior staining, drying or other treatments which might disturb the structure. The discussion in Sect. 3 furthermore demonstrates that equipment using an ordinary X-ray source is sufficient to analyze latex particles up to diameters of more than 150 nm. In future, this technique will certainly be applied more often to the investigation of latexes and become a standard tool for the analysis of these systems.

Acknowledgment. The authors are indebted to Prof. W. Ruland for helpful discussions. Financial support from the Deutsche Forschungsgemeinschaft, by the Geschäftsbereich Kautschuk, Bayer AG, by the AIF (project 9749), and by the Bundesministerium für Bildung und Forschung is gratefully acknowledged.

6
References

1. Guinier A, Fournet G (1955) Small-Angle Scattering of X-Rays, Wiley, New York
2. Glatter O, Kratky O (ed) (1982) Small Angle X-Ray Scattering, Academic Press, London
3. Feigin LA, Svergun DI (1987) Structure Analysis by Small-Angle X-Ray and Neutron Scattering, Plenum Press, New York
4. Lindner P, Zemb Th, (eds) (1991) Neutron, X-Ray and Light Scattering: Introduction to an Investigative Tool for Colloidal and Polymeric Systems, North Holland, Amsterdam

5. Hunter RJ (1989) Foundations of Colloid Science, Clarendon Press, Oxford
6. Backus RC, Williams RC (1948) J Appl Phys 19:1186; (1949) J Appl Phys 20:224
7. Gerould CH (1950) J Appl Phys 21:183
8. Yudowitch KL (1949) J Appl Phys 20:174; (1951) J Appl Phys 22:214
9. Leonard, Jr.BR, Anderegg JW, Kaesberg P, Beeman WW (1952) J Appl Phys 23:152
10. Danielson WE, Shenfil L, DuMond JWM (1952) J Appl Phys 23:860
11. Henke BL, DuMond JWM (1955) J Appl Phys 26:903
12. Motzkus F (1959) Acta Cryst 12:773
13. Vanderhoff JW (1969) In: Ham GE (ed) Vinyl Polymerization, Vol. 1, Part II, Dekker, New York
14. Higgins JS, Benoit HC (1994) Polymers and Neutron Scattering, Clarendon Press, Oxford
15. Ottewill RH (1990), In: Candau F, Ottewill RH, (eds) Scientific Methods for the Study of Polymer Colloids and Their Applications, Chapt 15, Kluwer Academic Publishers, Dordrecht
16. Goodwin JW, Ottewill RH, Harris NM, Tabony J(1980), J Colloid Interface Sci 78:253
17. Fisher LW, Polder SM, O'Reilly JM, Ramakrishnan V, Wignall GD (1988) J Colloid Interface Sci 123:24
18. Wai M, Gelman RA, Hoerl RH, Wignall GD (1987) Polymer 28:918
19. Hergeth WD, Bittrich HJ, Eichhorn F, Schlenker S, Schmutzler K, Steinau U-J (1989) Polymer 60:1913
20. Yang SI, Klein A, Sperling LH, Casassa EF (1990) Macromolecules 23:4582
21. Yang SI, Klein A, Sperling LH (1989) J Polym Sci Polym Phys 27:1649
22. Ottewill RH (1990) Faraday Disc Chem Soc 90:1
23. Mills MF, Gilbert RG, Napper DH, Rennie AR, Ottewill RH (1993) Macromolecules 26:3553
24. Linné MA, Klein A, Sperling LH, Wignall GD (1988) J Macromol Sci Phys B27:181
25. Bootle GA, Lye J, Ottewill RH (1990) Makromol Chem Macromol Symp 35/36:291
26. Ottewill RH (1992) Progr Colloid Polym Sci 88:49
27. Ottewill RH, Cole SJ, Waters JA (1995) Macromol Symp 92:97
28. Harris NM, Ottewill RH, White JW (1983), In: Ottewill RH, Rochester CH, Smith AC (eds) Adsorption from Solution, Academic Press, London
29. Ottewill RH,. Sinagra E, MacDonald IP, Marsh JF, Heenan RK (1992) Colloid Polym Sci 270:602 and further references given therein
30. Fleer GJ, Cohen Stuart MA, Scheutjens JMHM, Cosgrove T, Vincent B (1993) Polymers at Interfaces, Chapman & Hall, London
31. BushukW, Benoit H (1958) Can J Chem 36:1616
32. Leng M, Benoit H (1961) J Chim Phys 58:480
33. Benoit H (1966) Ber Bunsenges Phys Chem 70:286
34. Stuhrmann HB, Kirste RG (1965) Z Phys NF 46:247
35. Stuhrmann HB, Kirste RG (1967) Z Phys NF 56:334
36. Kirste RG, Stuhrmann HB (1967) Z Phys NF 56:338
37. Stuhrmann HB, Fuess H (1976) Acta Cryst A32:67
38. Luzatti V, TardieuA, Mateu L, Stuhrmann HB (1976) J Mol Biol 101:115
39. Luzatti V, Tardieu A (1980) Ann Rev Biophys Bioeng 9:1
40. Philipse AP, Smits C, Vrij A (1989) J Colloid Interf. Sci 129:335
41. Paradies H (1980) J Phys Chem 84:599
42. Lindner P, May RP, Timmins PA (1992) Physica B 180&181:967
43. Beyer D, Lebek W, Hergeth W-D, Schmutzler K (1990) Colloid Polym Sci 268:744
44. Hergeth W-D, Schmutzler K, Wartewig S (1990) Makromol Chem Macromol Symp 31:123
45. Grunder R, Kim YS, Ballauff M, Kranz D, Müller H-G (1991) Angew Chem Int Ed 30:1650
46. Grunder R, Urban G, Ballauff M (1993) Colloid Polym Sci 271:563

47. Dingenouts N, Ballauff M (1993) Acta Polymerica 44:178
48. Dingenouts N, Kim YS, Ballauff M (1994) Makromol Chem Rapid Commun 15:613
49. Dingenouts N, Kim YS, Ballauff M (1994) Colloid Polym Sci 272:1380
50. Dingenouts B, Pulina T, Ballauff M (1994) Macromolecules 27:6133
51. Ballauff M, Bolze J, Dingenouts N, Hickl P, Pötschke D (1996) Macromol Chem 197:3043
52. Bolze J, Ballauff M (1995) Macromolecules 28:7429
53. Bolze J, Hörner KD, Ballauff M (1996) Langmuir 12:2906
54. Bolze J, Hörner KD, Ballauff M (1996) Colloid Polym Sci 274:1099
55 Bolze J, Hörner KD, Ballauff M (1997) Langmuir 13:2960
56. Harrison SC (1969) J Mol Biol 42:457
57. Feigin LA, Sholer I (1975) Sov Phys Crystallogr 20:302
58. Kawaguchi T, Hamanaka T (1992) J Appl Cryst 25:778; Kawaguchi T (1995) J Appl Cryst 28:424 and further references given there
59. Hickl P, Ballauff M, Jada A (1996) Macromolecules 29:4006
60. Hickl P, Ballauff M (1997) Physica A 235:238
61. Thies M, Hinze U, Paradies HH (1995) Colloids Surf 101:261
62. Russel WB, Saville DA, Schowalter WR (1989) Colloidal Dispersions, Cambridge University Press, Cambridge
63. D'Aguanno B, Klein R (1996) In: Brown W (ed) Light Scattering, Principles and Applications, Clarendon Press, Oxford
64. Apfel U, Grunder R, Ballauff M (1994) Coll Pol Sci 272:820
65. Apfel U, Hörner KD, Ballauff M (1995) Langmuir 11:3401
66. Hansen JP, McDonald IR (1986) Theory of Simple Liquids, Second Edition, Academic Press, London
67. Vrij A (1979) J Chem Phys 71:3267
68. van Beurten P, Vrij A (1981) J Chem Phys 74:2744
69. Pedersen JS (1995) In: Brumberger H (ed) Modern Aspects of Small-Angle Scattering. Kluwer Academic Publishers, Dordrecht
70. Kratky O (1954) Z Elektrochem 58:49
71. Kratky O (1982) In: Small Angle X-ray Scattering, Glatter O and Kratky O (ed), London, Academic Press, chapter 3
72. Glatter O (1982) In: Small Angle X-ray Scattering, Glatter O and Kratky O (ed) London, Academic Press, chapter 4
73. Dingenouts N, Ballauff M (1998) Acta Polymerica 49:178
74. Bösecke P, Diat O, Rasmussen B (1995) Rev Sci Instrum 66:1636
75. Stabinger H and Kratky O (1978) Makromol Chem 179:1655
76. Glatter O and Zipper P (1975) Acta Physica Austr 43:307
77. Beniaminy I and Deutsch M (1980) Comp Phys Comm 21:271
78. Strobl GR (1970) Acta Cryst A26:367
79. Schelten J and Hossfeld F (1971) J Appl Cryst 4:210
80. Lake JA (1967) Acta Cryst 23:191
81. Weiss A, Ballauff M, in preparation
82. Müller K (1982) In: Small Angle X-ray Scattering, Glatter O and Kratky O (ed) London, Academic Press
83. Pollizi S, Stribeck N, Zachmann HG and Bordeianu R (1989) Colloid Polym. Sci. 267:281
84. Weast RC (ed) (1982) CRC Handbook of Chemistry and Physics Boca Raton, CRC Press
85. Rathje J, Ruland W (1976) Colloid Polym Sci 254:358; Ruland W (1977) Pure Appl Chem 49:905; Wiegand W, Ruland W (1979) Progr Colloid Polym Sci 66:355
86. Koberstein JT, Morra B and Stein RS (1980) J Appl Cryst 13:34
87. Müller HG (1989) Colloid Polym Sci 267:1113
88. Bösecke P (1992) Rev Sci Instr 63:438
89. Megens M, van Kats CM, Bösecke P, Vos WL (1997) Langmuir 13:6120

90. Bonse U, Hart M (1965) Appl Phys Lett 7:238; Bonse U, Hart M (1966) Z Phys 189:151
91. Chu B, Li Y, Gao T (1992) Rev Sci Instr 63:4128
92. Koga T, Hart M, Hashimoto T (1996) J Appl Cryst 29:318
93. Reus V, Belloni L, Zemb T, Lutterbach N, Versmold H (1997) J Phys II France 7:603
94. Reus V, Belloni L, Zemb T, Lutterbach N, Versmold H (1995) J Chim Phys 92:1233
95. Konishi T, Ise N, Matsuoka H, Yamaoka H, Sogami IS, Yoshiyama T (1995) Phys Rev B 51:3914
96. Diat O, Bösecke P, Ferrero, C, Freund AK, Lambard J, Heintzmann R (1995) Nucl Instr and Meth. Phys Res A356:566
97. Gilbert RG (1995) Emulsion Polymerization, A Mechanistic Approach, Academic Press, London
98. Binanco H, Narkis M, Cohen Y (1996) J Polym Sci Polym Phys 34:2775
99. Casassa EF (1967) J Polym Sci Polym Lett Ed 5:773
100. Casassa EF, Tagami Y (1969) Macromolecules 2:14
101. de Gennes PG (1981) Macromolecules 14:1637
102. de Gennes PG (1987) Adv Colloid Interf Sci 27:189
103. Croxton CA, Mills MF, Gilbert RG, Napper DH (1993) Macromolecules 26:3563
104. Ronca G (1987) J Appl Polym Sci 33:2623
105. Scheutjens IMHM, Fleer GJ (1982) Adv Colloid Interf Sci 16:361
106. Vincent B (1990) Colloid Surf 50:241
107. Brazhnik PK, Freed KJ, Tang H (1994) J Chem Phys 101:9143
108. Evers OA, Ley G, Hädicke E (1993) Macromolecules 26:2885
109. Evers O, Hädicke E, Ley G (1994) Colloids Surf A 90:135
110. Chang HS, Chen SA (1987) Makromol. Chem Rapid Commun 8:297
111. Dabdub D, Klein A, Sperling LH, (1992) J Polym Sci Polym Phys 30:787
112. Kayes JB, (1976) J Colloid Interf Sci 56:426
113. Huldén M, Kronberg B (1994) J Coat Technol 66:67
114. Kronberg B, Lindström M, Stenius P (1986) In: Phenomena in Mixed Surfactant Systems, Scamehorn JF (ed), A.C. S. Symposium Series 311, American Chemical Society, Washington D.C.
115. Zemb T, Charpin P (1985) J Physique 46:249

Editor: Prof. A. Abe
Received: January 1998

Aliphatic-Cycloaliphatic Epoxy Compounds and Polymers

Anatoli E. Batog[1], Ivan P. Pet'ko[1], Piotr Penczek[2]

[1] Ukrainian State Scientific-Research Institute of Plastics (UkrGos NII PlastMass) Prospekt Il'icha 95, 340059 Donetsk, Ukraine
[2] Industrial Chemistry Research Institute (ICRI) ul. Rydygiera 8, 01–793 Warsaw, Poland

Aliphatic-cycloaliphatic epoxy compounds (ACECs) contain different epoxy groups in the molecule: glycidyl, i.e. 2,3-epoxypropyl groups, and cycloaliphatic, i.e. 1,2-epoxycyclopentane or 1,2-epoxycyclohexane rings. The epoxidation of double bonds in cycloolefins was carried out using aqueous peracetic acid (PAA). The synthesis of aqueous PAA is described. The kinetics of the epoxidation of cycloolefins with PAA is reviewed. Various ACECs, their synthesis and properties are given. The ACECs involve, among others, glycidyl ethers and esters, glycidyloxyphenyl derivatives, cyanoethylated compounds, imide ring-containing products and aliphatic-cycloaliphatic epoxy oligomers. The ACECs were crosslinked with acid anhydrides, amines and phenol-formaldehyde oligomers. The effect of the structure of the ACECs and of the curing agents on the properties of the cast crosslinked polymers and the composites with the ACEC based polymer matrix is discussed. The advantageous effect of the addition of ACECs on the properties of cycloaliphatic epoxy compounds (CECs) is presented. It is assumed that the different reactivity of glycidyl and cycloaliphatic epoxy groups results in an increased structural regularity of the crosslinked ACECs, thus improving the mechanical and thermal properties. The effect of modification of epoxy compounds and resins with ACECs on their properties is considered.

keywords: Epoxy resins, Cycloaliphatic epoxy compounds, Glycidyl groups, Peracetic acid, Crosslinking of epoxy compounds, Acid anhydride curing agents, Amine curing agents, Phenol-formaldehyde resins, Composites, Kinetics of epoxidation, Structure regularity

Advances in Polymer Science, Vol.144
© Springer-Verlag Berlin Heidelberg 1999

List of Symbols and Abbreviations

ACE   aliphatic-cycloaliphatic epoxy compound
CEC   cycloaliphatic epoxy compound

CODDO   cyclooctadiene dioxide
COL   cycloolefin
DCP   dicyclopentadiene
DCPDO   dicyclopentadiene dioxide
ECH   epichlorohydrin
IR   infrared
PA   peracid
PAA   peracetic acid
PAFO   phenol-aniline-formaldehyde oligomer (resin)
PFO   phenol-formaldehyde oligomer (resin)
PVC   poly(vinyl chloride)
THI   tetrahydroindene
THIDO   tetrahydroindene dioxide
TM   heat deflection temperature (Martens)
UP-XXX   Ukrainian polymers, i.e. the epoxy resins (compounds) manufactured by the Ukrainian State Scientific-Research Institute of Plastics
UP-XXXT as above, triepoxide resins (compounds)
UV   ultraviolet

1
Introduction

Cycloaliphatic diepoxy compounds (CECs) were first synthesized in 1950 and were used in the early 1960s as UV resistant plastics for the outdoor electric insulation [1–3].

Compounds which contain epoxy groups in the cycloaliphatic ring as well as in the pendant glycidyl (i.e. 2,3-epoxypropyl) groups, so-called aliphatic-cycloaliphatic epoxy compounds (ACECs), have been less known. The methods of industrial synthesis of ACECs and the future areas of application were scarcely described until the first articles appeared in the 1970s [4]. Then many new ACECs were prepared and the methods of preparation were described. The syntheses were followed by the investigation of the properties of crosslinked ACECs and the consideration of possible applications.

The simplest ACECs are glycidyl ethers (I, Scheme 1) or esters (II, Scheme 2) linked to an epoxycyclohexane ring.

The most important feature of ACECs is the different reactivity of the cycloaliphatic epoxy group and the glycidyl epoxy group with various curing agents. This property affects some important properties of ACECs.

$$O \diamondsuit\!\!\!\!\bigcirc -CH_2-O-CH_2-CH-CH_2 \atop \diagdown O \diagup$$ (I)

Scheme 1

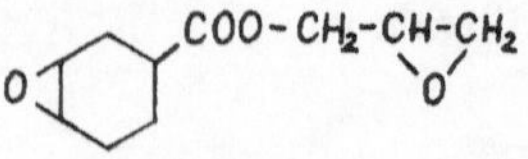

$$\text{O}\underset{}{\overset{}{\bigotimes}}\!\!-\!\!COO\!-\!CH_2\!-\!CH\!-\!CH_2 \qquad\qquad (II)$$

Scheme 2

In this article, experimental results concerning the synthesis of ACECs, their properties, as well as the properties of ACECs-based crosslinked polymers and composites, are reviewed. The kinetics of the synthesis and curing of ACECs is considered only to the extent needed for the evaluation of the reactivity of different epoxy groups in ACECs and for the understanding of the formation processes of the crosslinked ACEC polymers.

2
Epoxidation of Cycloolefins Containing Unsaturated Substituents

2.1
General Considerations

Cycloolefins (COLs) with unsaturated substituents are used as starting materials for the synthesis of ACECs. The epoxidation of COLs and of the COL unsaturation in particular with organic peracids (PAs) is the most important step of the synthesis.

The kinetic measurements have confirmed that the epoxidation of COLs with PAs is a second order reaction. The reaction rate depends on the concentration of the COL and PA as well on the nature of solvent [5–8] and its proton donor properties [7]. The reaction rate is influenced by the dielectric constant of the reaction medium, although there has been no salt effect and no autocatalysis observed [7, 9, 10]. The kinetic and thermodynamic data are in accordance with the mechanism of the COLs epoxidation with PAs which was suggested by Bartlett [11]. The reaction rate of the epoxidation of COLs with PAs is strongly af-

Table 1. Epoxidation of 3 -substituted cyclohexene derivatives with perbenzoic acid at 40 °C [12]

Substituent	Relative rate of epoxidation		*cis/trans* epoxy ratio (%)	
	in benzene	in chloroform	in benzene	in chloroform
H	100	100	–	–
methyl	50	44	50:50	60:40
isopropyl	38	19	18:82	4:96
tert–butyl	31	16	10:90	10:90
phenyl	29	11	5:95	5:95

fected by the nucleophilic properties of the double bond: electron donor substituents increase the reaction rate, whereas the electron acceptor substituents cause a decrease in the reaction rate [12]. The effect of substituents in the cyclohexene ring on the epoxidation rate is presented in Table 1.

Similar results were obtained with the epoxidation of bicyclic olefins [13].

2.2
Preparation of Solutions of Peracetic Acid

Epoxidation of unsaturated compounds is usually carried out in the industry by means of anhydrous solutions of peracetic acid (PAA) in ethyl acetate. The PAA solutions are manufactured by the oxidation of acetaldehyde with ozonized oxygen.

We applied aqueous solution of PAA for safety reasons. The manufacturing of aqueous PAA involves the reaction of 40–60% commercial H_2O_2 with acetic acid in the presence of a mixture of strong mineral acids with a simultaneous removal of the PAA being formed. The PAA is distilled off from the reaction mixture in vacuo as an azeotrope with water. The temperature of the reaction mixture is kept below 55 °C. The pressure is 0.006–0.009 MPa. The yield of PAA is up to 98% (calculated on H_2O_2). The addition rate of the equimolar H_2O_2/CH_3COOH mixture should correspond to the rate of PAA removal.

The composition of the distillate depends mainly on the concentration of the introduced aqueous H_2O_2. The product contains 35–55% PAA, 1–3% H_2O_2, 8–20% CH_3COOH and 34–45% H_2O. The advantage of the PAA solution lies in the relatively low concentration of acetic acid and in the absence of strong mineral acids [12].

The process of manufacturing of aqueous PAA described above can easily be made continuous. The PAA content can be varied by changing the concentration of the input H_2O_2. Our many years experience confirmed the feasibility of the technology.

It has been found that the aqueous solutions of PAA which do not contain mineral acids exhibit a satisfactory stability in the PAA concentration range of 16–55% at a pH of 2.2–2.4. No more than 5% PAA is decomposed in 4 h at 20 °C. However, the stability is considerably decreased at a pH of 3.6–4.6, being the most favorable for the epoxidation of COLs. This pH range is achieved by an addition of buffer salts, e.g. sodium acetate or bicarbonate. When the pH value is thus increased, up to 25% PAA is decomposed in 4 h at the same temperature.

The addition of 0.05% Na tripolyphosphate as stabilizer substantially decreases the PAA decomposition rate. Thus, if using stabilized and buffered aqueous PAA at pH 3.6–4.6 for the investigation of the COLs epoxidation, the decomposition of PAA can be neglected.

2.3
Epoxidation of Cycloolefins with Aqueous Peracetic Acid.
Kinetics of Epoxidation

Application of aqueous PAA for the epoxidation of vegetable oils to manufacture epoxy stabilizers for PVC has become a common process [14]. However, there is less information about the use of aqueous PAA for the epoxidation of COLs and very little data on the kinetics of such reaction are available.

It has been found [15] that the epoxidation rate with various peracids R-C(=O) OOH increases in the order R: CH_3- $<C_6H_5CH_2$- $<p$-CH_3-O-C_6H_4-$<C_6H_5$- $<\alpha$-and β-$C_{10}H_7$- $<p$- and m- O_2N-C_6H_4-. Thus, PAA belongs to the group of less efficient epoxidizing agents.

Reactivity of COLs in the epoxidation reaction depends on the structure of the cycloolefin [16]. A series of model COLs was used that represent COLs that have been used for the manufacture of diepoxy compounds((Tables 2 and 3).

The epoxidation kinetics was investigated in acetic acid in the temperature range of 10–40 °C. The dilution with acetic acid should not affect the epoxidation rate because PAA in such systems is almost completely associated with acetic acid even if not additionally diluted with acetic acid. It was found that PAA was consumed exclusively for epoxidation within the given temperature range and time.

The data given in Table 2 show that the ester bonds in the COLs **III** and **IV** as well as the cyclic imide group in **XIII** considerably decrease the epoxidation rate in comparison with non-substituted cyclohexene as well as with cyclohexene compounds bearing substituents other than ester groups. It may be due to the electron acceptor character of the ester group. The presence of two groups (**IV**) and two carbonyls in the cyclic imide **XIII** decreases the reaction rate even more.

A decrease in the reaction rate was also observed in case of the compounds **VIII, IX, X** and **XI** which contain the dioxane ring.

As far as the dioxolane ring-containing compounds are concerned, no conclusions about the influence of the substituents on the reaction rate can be drawn. Thus, the epoxidation rate constant k of the compounds **X, VI** and **VII** differs only a little from that of cyclohexene. The introduction of two methyl groups into the cyclohexene ring (**VI**) does not increase the reaction rate although the methyl group has electron-donor properties.

High values of the epoxidation rate of 2-furylmethyl 3-cyclohexenecarboxylate (**XII**) cannot be explained by the induction effect of the substituent. Apart from this effect, which is transmitted through the δ-linkages and through the space inside and around the rings (the electrostatic field effect), there are steric effects and the specific solvatation of the **XII** molecule with acetic acid.

It should be mentioned that the k value is almost constant in course of the epoxidation of tetrahydroindene (THI, **XV**), up to high conversion values.This phenomenon, as well as the fact that the activation parameters of THI and cyclohexene have similar values illustrate the approximately equal reactivity of both double bonds in THI.

Table 2. Reaction rate constants and the activation energy parameters of the epoxidation of COLs with PAA; the starting concentrations $[PAA]_0 = [COL]_0 = 0.6 - 1.4$ mol kg^{-1}

Compound	Scheme	$k \cdot 10^4$ (kg mol^{-1}s^{-1})				E (kJmol^{-1})	ΔH (kJmol^{-1})	$-\Delta S$ (J.mol^{-1}K^{-1})
		10 °C	20 °C	30 °C	40 °C			
III	3	1.40	3.59	7.90	18.4	65.1	62.6	88.5
IV	4	1.03	2.10	4.48	9.44	55.9	53.4	122.1
V	5	3.51	8.40	–	35.6	55.5	53.0	111.0
VI	6	2.99	7.12	–	–	–	–	–
VII	7	–	12.5	25.4	58.3	57.9	55.5	99.8
VIII	8	1.70	3.51	–	–	–	–	–
IX	9	1.62	3.14	7.34	13.1	49.7	47.3	139.0
X	10	–	4.45	12.6	30.3	75.5	70.1	57.5
XI	11	2.20	4.16	10.4	–	–	–	–
XII	12	4.18	8.81	22.3	44.7	57.4	54.9	106.1
XIII	13	–	1.89	4.75	11.4	68.0	65.6	78.7
XIV	14	2.68	5.99	13.8	–	59.5	57.1	99.6
XV	15	3.85	9.45	22.7	–	59.9	57.4	106.2
XVI	16	3.66	8.93	21.4	–	62.8	60.3	85.2

Table 3. Properties of COLs given in Table 2

| Compound (Table 2) | Formula | Elementary analysis (%) | | | | | | | | Boiling temp. (°C) | Melting temp. (°C/Pa) | Density d_4^{20} | Refracion index n_D^{20} |
| | | Calculated | | | | Found | | | | | | | |
		Bromine Number	C	H	N	Bromine Number	C	H	N				
III	$C_8H_{12}O_2$	114.14	68.65	8.63	–	107.31	68.12	8.93	–	56–57/700	–	1.1118	1.4637
IV	$C_{10}H_{14}O_4$	80.70	60.60	7.12	–	78.58	60.01	7.44	–	131–132/800	–	1.0804	1.4721
V	$C_9H_{14}O_2$	103.76	70.10	9.15	–	102.28	69.98	9.56	–	88–90/700	–	1.0473	1.4820
VI	$C_{11}H_{18}0_2$	87.68	72.49	9.95	–	86.83	72.01	9.39	–	139–140/700	–	1.0497	1.4805
VII	$C_{13}H_{20}O_4$	66.51	64.97	8.39	–	64.98	64.76	8.58	–	–	75–76	–	–
VIII	$C_9H_{14}O_2$	103.63	70.10	9.15	–	103.11	69.98	9.31	–	82–83/500	–	1.0051	1.4880
IX	$C_{14}H_{22}O_2$	71.88	75.63	9.97	–	71.52	75.12	9.83	–	135–136/130	–	1.0891	1.5109
X	$C_{13}H_{16}O_3$	72.55	70.89	7.32	–	71.99	70.89	7.32	–	150–151/800	–	1.1406	1.5183
XI	$C_{15}H_{18}0_2$	69.38	78.83	7.88	–	68.90	78.03	7.91	–	–	39	–	–
XII	$C_{12}H_{14}O_3$	77.48	69.88	6.84	–	77.02	69.12	6.68	–	120–121/800	–	1.1281	1.4848
XIII	$C_{14}H_{13}O_2$	70.31	73.99	5.79	6.16	69.94	73.65	5.44	5.98	–	116	–	–

The activation parameters of the epoxidation reaction were calculated from the temperature dependence of the k values (Table 2). The E values of the compounds **II–XVI** are in the range of 50-72 kJ mol^{-1} (mostly 55–60 kJ mol^{-1}) and the Δ S values exhibit highly negative values of 85–139 J mol^{-1}K^{-1}.

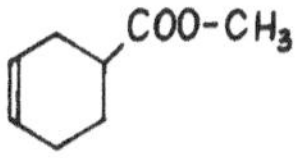

(III)

Scheme 3

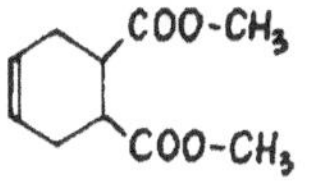

(IV)

Scheme 4

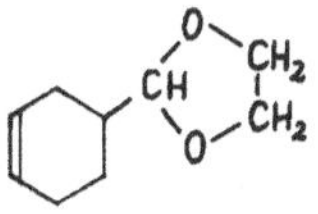

(V)

Scheme 5

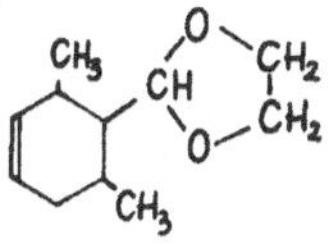

(VI)

Scheme 6

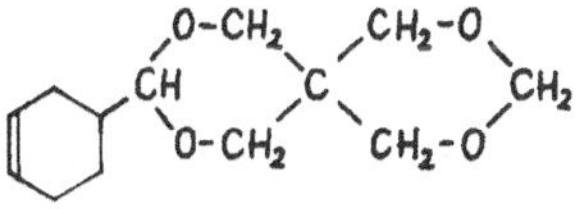

(VII)

Scheme 7

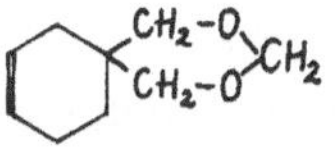

(VIII)

Scheme 8

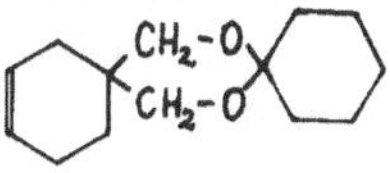

(IX)

Scheme 9

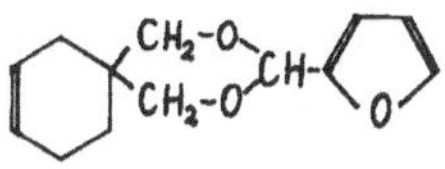

(X)

Scheme 10

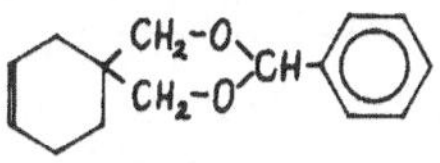

(XI)

Scheme 11

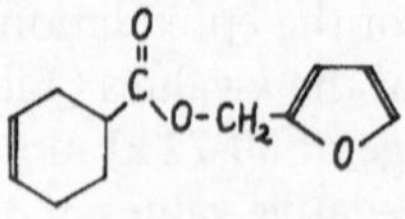

(XII)

Scheme 12

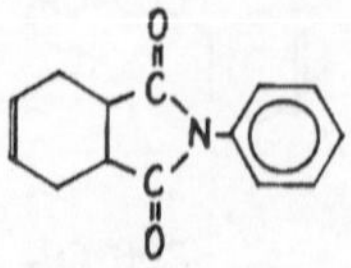

(XIII)

Scheme 13

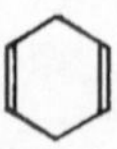

(XIV)

Scheme 14

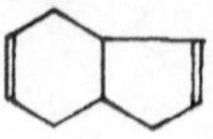

(XV)

Scheme 15

(XVI)

Scheme 16

There exists a linear dependence between the enthalpy and entropy values (Fig. 1). The slope of the straight line, which was considered as the isokinetic temperature, amounts to 3K. It was calculated by the least squares method. Similar values of the isokinetic temperature were obtained by means of the method of intersection of the Arrhenius lines.

Consequently, the occurence of the compensation effect serves as evidence of the generality of the epoxidation mechanism of the investigated COLs. We believe that the proximity of the epoxidation temperature of the investigated COLs and the isokinetic temperature is the conclusive factor which explains the minor influence of the structure of the substituents on the epoxidation rate.

2.4
Stability of Epoxy Groups During the Epoxidation with Aqueous Peracetic Acid

Possibility of the synthesis of complex ACECs can be evaluated only on the grounds of the analysis of kinetic parameters of the epoxidation of cyclohexene derivatives having glycidyl ether or ester substituents. Moreover, the stability of the glycidyl epoxy groups in the reaction medium should be taken into account. Epoxidation rate of the cyclohexene ring with glycidyl containing substituents (Table 4) is a little lower than that of cyclohexene and two to three times higher

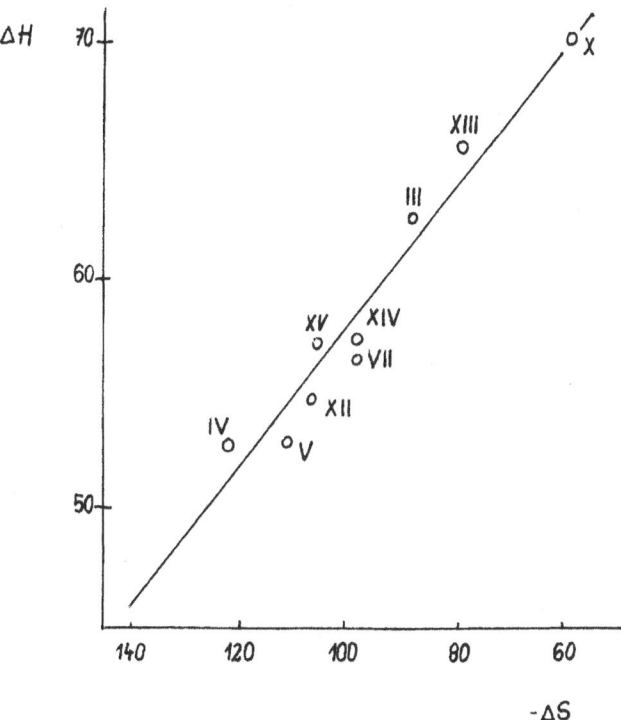

Fig. 1. Compensation effect in the epoxidation reaction of COLs with PAA. The Roman numerals correspond to the numbers of compounds in Tables 2 and 3 and in the text

in comparison with methyl cyclohexene carboxylate (Table 2). Although the glycidyl substituents decrease the epoxidation rate of the cyclohexene ring, that rate remains close to that given for most substituted COLs presented in Table 1.

As far as the stability of the glycidyl group in ACECs in the epoxidation conditions of the cyclohexene ring is concerned, it has been found that glycidyl cyclohexene carboxylate (**XV**) and cyclohexenemethyl glycidyl ether (**XVI**) do not

Table 4. Epoxidation rate constants (k) and activation parameters of the reaction of glycidyl derivatives of cyclohexene with PAA

Com-pound	Scheme	$k \cdot 10^4 (\text{kg mol}^{-1}\text{s}^{-1})$				E	ΔH	$-\Delta S$
		10 °C	20 °C	30 °C	40 °C	(kJ mol^{-1})	(kJ mol^{-1})	(kJ mol^{-1}K^{-1})
XVII	17	2.73	4.90	–	18.71	47.76	45.29	140.77
XVIII	18	4.52	9.49	28.69	–	65.01	62.54	45.34
XIX	19	2.68	4.99	8.60	–	41.40	38.93	162.35

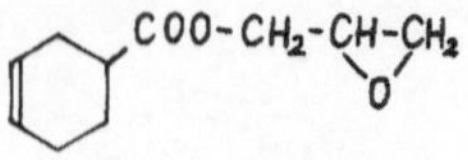

(XVII)

Scheme 17

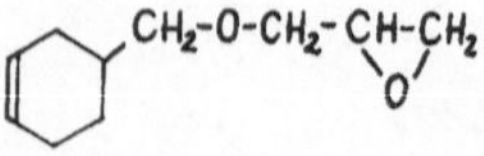

(XVIII)

Scheme 18

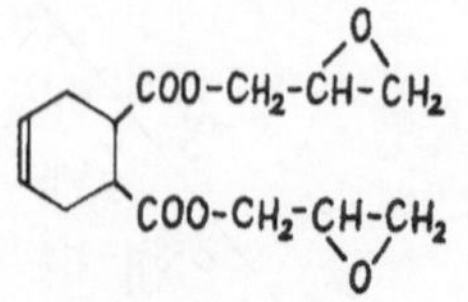

(XIX)

Scheme 19

Scheme 20

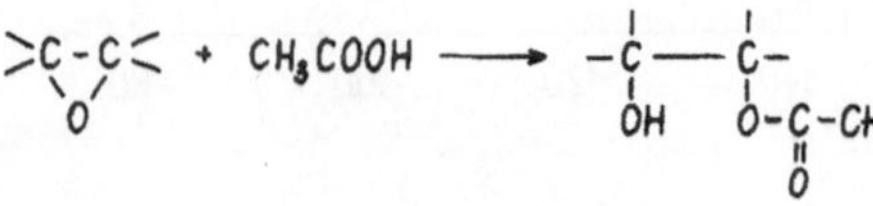

Scheme 21

react with 50% and 100% acetic acid at temperatures up to 65 °C. Consequently, various substituted COLs can be epoxidized in the selected optimum conditions.

The effect of the reaction medium on the selectivity of epoxidation is best characterized by the ratio of the epoxidation rate and the epoxy ring opening. It has been found [17] that a decrease in water content in PAA increases the selectivity of epoxidation. Water and acetic acid which are contained in PAA can react with the formed epoxy ring as shown below (Schemes 20 and 21).

If anhydrous PAA is applied, the reaction with acetic acid is suppressed.

We investigated the side reactions of CECs with acetic acid and its aqueous solutions at 10–40 °C. The reaction mixtures served as models of the epoxidation systems. The course of the reactions was followed by the determination of epoxy and carboxyl groups in the reaction mixture. Equimolar ratios of the reagents were applied [18].

The epoxy ring opening with acetic acid occurs at a temperature as low as 40 °C. The reaction rate depends on the structure of the epoxy compound (Fig. 2). Electron donor substituents ($-COOCH_3$) decrease the electron density of epoxy groups, resulting in a decrease in the reaction rate with the carboxylic acid. This influence is particularly strong in the case of the esters **XX** and **XXI**. Such compounds are the most stable ones in epoxidation conditions.

The epoxy compounds with acetal rings exhibit lower stability in the acidic medium. The reaction rate of the epoxy groups with acetic acid is particularly accel-

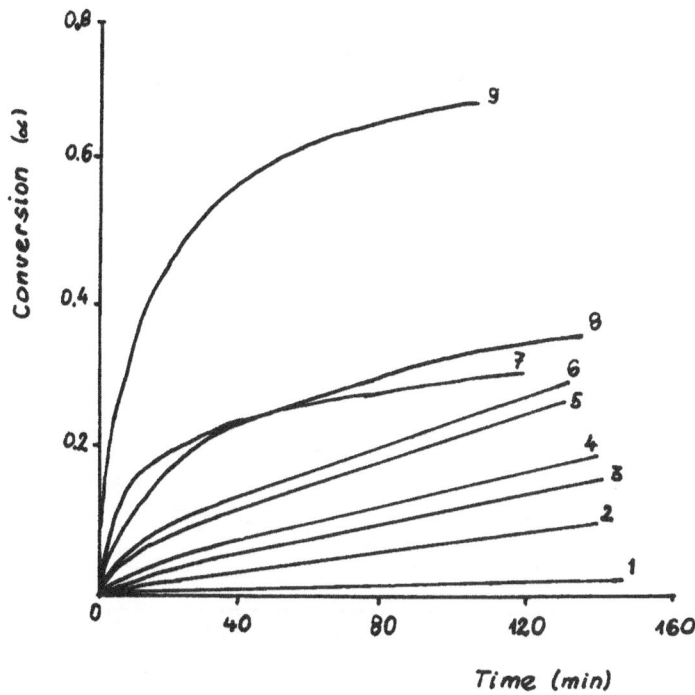

Fig. 2. Conversion (α) of oxirane ring in CECs in 100% acetic acid. Molar ratio of epoxide: acid=1:1; 1 – **XX** (Scheme 22), 40 °C; 2 – **XX**, 60 °C; 3 – **XXI** (Scheme 23), 40 °C; 4 – **XXII** (CODDO), Scheme 24), 40 °C; 5 – **XXIII** (THIDO, Scheme 25), 40 °C; 6 – **XXIV** (Scheme 26), 40 °C; 7 – **XXV** (Scheme 27), 40 °C; 8 – **XXVI** (Scheme 28), 40 °C; 9 – **XXV**, 60 °C

erated by increasing temperature (Fig. 2, curves 7 and 9). 1,5-Cyclooctadiene dioxide (CODDO) and tetrahydroindene dioxide (THIDO) take an intermediate position between the epoxycyclohexane derivatives with ester and acetal groups as far as the reaction rate of the epoxy group with acetic acid is concerned.

It has been found that the character of the substituent effect on reactivity does not change when various concentrations of the aqueous CH_3COOH are applied. An increase in the dissociation degree of acetic acid, which is caused by the increased dilution and temperature, results in an acceleration of the epoxy groups conversion (Fig. 3).

An addition of 5% CH_3COONa (NAC, calculated on CH_3COOH) increases the pH value of the reaction mixture up to 4.5. Consequently, the epoxy ring opening rate is considerably decreased (Fig. 4). At a temperature of 10–20 °C, the conversion of epoxy groups becomes negligible. In the presence of solvents, the epoxy groups conversion is almost completely stopped (Fig. 5). Solvation of the reacting groups with the solvent presumably results in an increase in the activation energy and, consequently, in a decrease in the acid/epoxy reaction rate

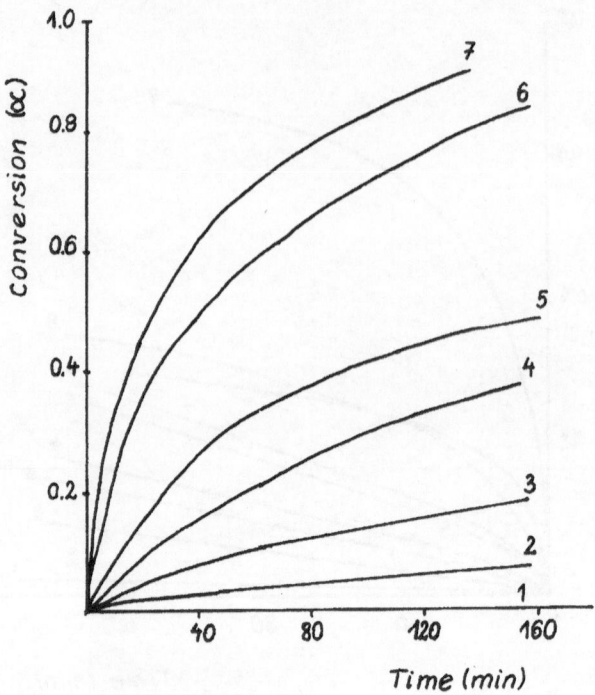

Fig. Stab XfQDDD(X)hemaqueeulatiofsitious concentratianterperatureanversiphgroots;2G$_4^{25}$ 100%20 %30%40 %32%40 %35%20 %36%40 %72%40 °C

Table 5. Properties of some epoxycyclohexane derivatives

Compound (No.,Scheme)	Formula	Oxirane oxygen content %		Density d_4^{25}	Refraction Index n_D^{25}	Boiling temp. (°C //Pa)	Yield (%)
		Calcd.	Found				
XXI, 23	$C_8H_{12}O_3$	10.26	10.06	1.0928	1.4625	112// 400	93
XX, 22	$C_{10}H_{14}O_5$	7.48	7.24	1.1033	1.4711	123//133	95
XXVII, 30	$C_{10}H_{14}O_3$	8.86	9.16	1.0877	1.4622	84//200	86
XXVIII, 31	$C_{14}H_{18}O_5$	6.02	5.70	1.1123	1.4943	100//70	90
XXV, 27	$C_9H_{14}O_3$	9.41	9.31	1.1424	1.4790	95//200	97
XXIX, 32	$C_{13}H_{16}O_4$	6.78	6.76	–	–	182//400	88
II, 2	$C_{10}H_{14}O_4$	16.16	16.03	1.2971	1.4805	136//266	84

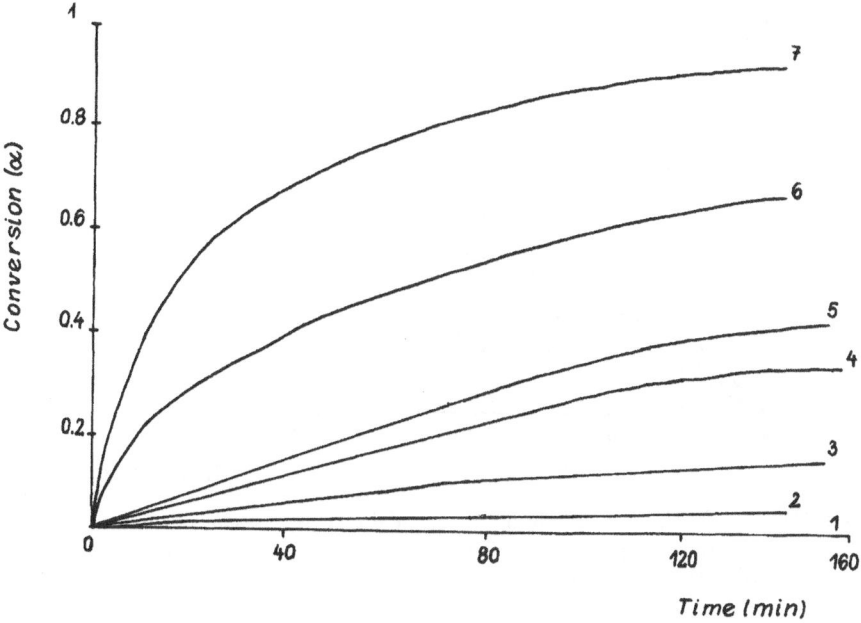

Fig. 4. Effect of Na acetate (NAC) on the stability of **XXV** in 50% acetic acid at various NAC concentrations and temperatures: 1 – 5.1%, 10 °C; 2 – 5.1%, 20 °C; 3 – 5.1%, 40 °C; 4 – 2.5%, 20 °C; 5 – 2.5%, 40 °C; 6 – 0%, 20 °C; 7 – 0%, 40 °C

(Fig. 6). The side reactions in the recommended conditions of the epoxidation reaction are thus of minor importance. It is also reflected by the high yields of the epoxy compounds (Table 5).

From the epoxidation kinetics of substituted cyclohexene derivatives with aqueous solutions of PAA one can conclude that the probability of side reactions

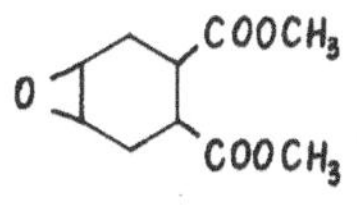

(XX)

Scheme 22

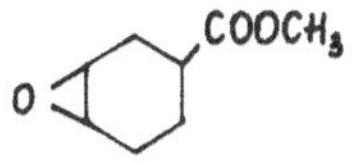

(XXI)

Scheme 23

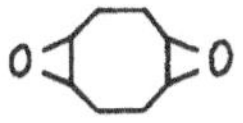

(XXII, CODDO)

Scheme 24

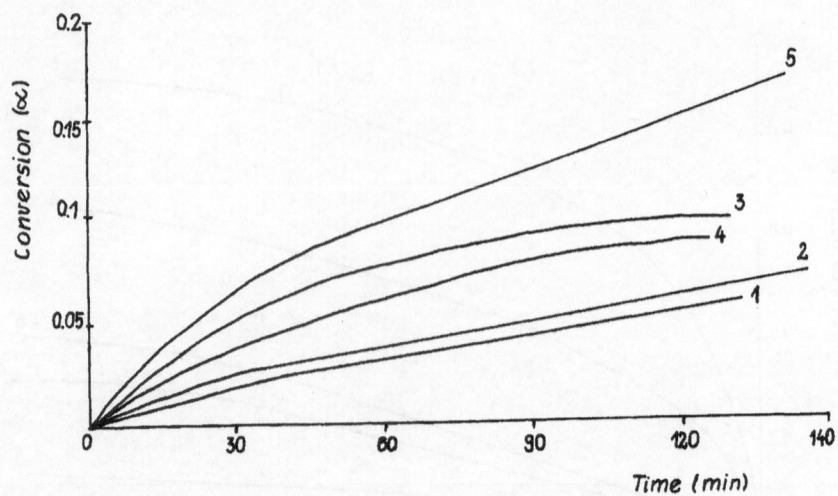

Fig. 5. Stability of CECs in the conditions of epoxidation with 55% aqueous PAA at 40%. Conversion (α) of epoxy groups: 1 – **XX**, in CHCl$_3$; 2 – **XX**, in toluene; 3 – **XX**, without solvent; 4 – **XXII** (CODDO), in toluene; 5 – **XXII**, without solvent

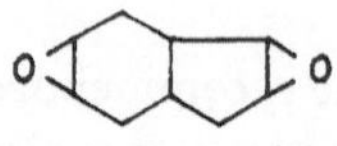

Scheme 25   (XXIII, THIDO)

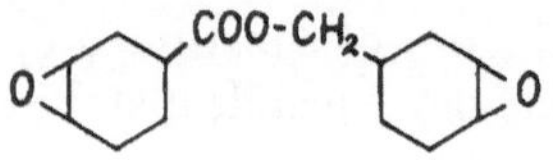

Scheme 26   (XXIV)

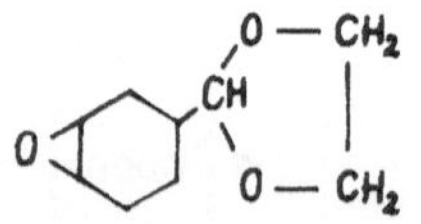

Scheme 27   (XXV)

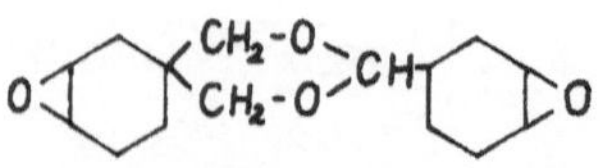

Scheme 28   (XXVI)

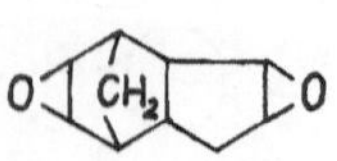

Scheme 29   (XXVII, DCPDO)

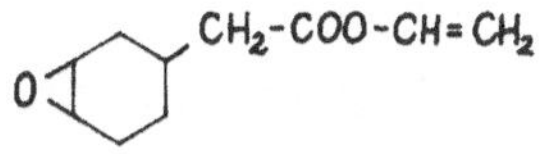

CH2-COO-CH=CH2
O

(XXVII)

Scheme 30

CH2-OOC-CH=CH2
O
CH2-OOC-CH=CH2

(XXVIII)

Scheme 31

CH2-O
CH2-O
CH
O
O

(XXVIII)

Scheme 32

PAA content (%)
30
20
10
0
2
4
6
8
Time (hrs)
1
2
3

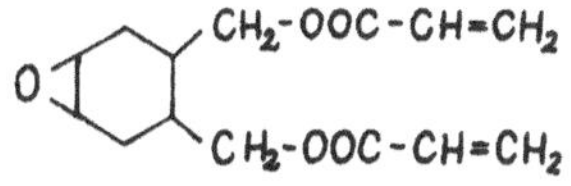

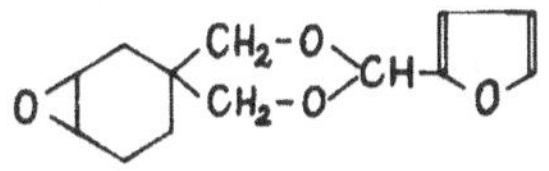

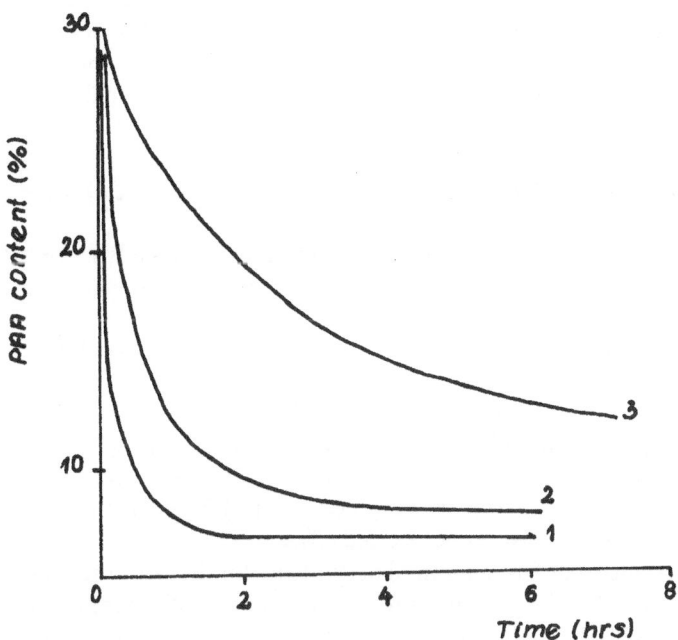

Fig. 6. Decrease in PAA content in the aqueous phase in course of the dicyclopentadiene (DCP) epoxidation in toluene at various temperatures: 1 – 50 °C; 2 – 40 °C; 3 – 20 °C

is very low provided the recommended conditions of the process are adhered to. High yields of the epoxy compounds and high epoxy groups content can be achieved, being even higher than that obtained with an anhydrous PAA solution.

All the above results show that the 40–50% aqueous solutions of PAA can be used as epoxidizing agents for a variety of substituted cyclohexene derivatives.

3
Synthesis of Aliphatic-Cycloaliphatic Epoxy Compounds (ACECs)

3.1
General Considerations

Heat resistant crosslinked polymers with excellent aging resistance, tracking and electric arc resistance and good dielectric properties over a broad temperature range are obtained by curing CECs. However, the application of CECs is limited due to the fragility which results from high crosslink density and rigid multicyclic structure of the polymer network. Thus, an increase in flexibility of crosslinked ACECs is of a considerable practical interest.

The conventional approach towards improving the impact resistance, based on the application of flexibilizing modifiers, hardeners and plasticizers, is not applicable. The increase in flexibility is accompanied by a decrease in heat resistance. Many mechanical, dielectric and other properties also deteriorate.

ACEC-based polymer networks exhibit an advantageous complex of properties including an elevated heat resistance and improved flexibility (Scheme 33).

3.2
Synthesis of Aliphatic-Cycloaliphatic Epoxy Compounds with Glycidyl Ether Groups

The synthesis of ACECs involves the well-known methods of preparation of CECs and glycidyl ethers. Cycloolefins with various functional groups are used as the starting materials.

Batzer [19] described a four-stage synthesis of an ACEC containing epoxycyclohexane ring and a pendant epoxyethyl (i.e. oxiranyl) group (Scheme 34). In this case, the chlorohydrin ether derivative of cyclohexene is first epoxidized with a peracid. Eventually, the dehydrochlorination step follows. The diepoxy compound thus prepared contains no more than 75% of the calculated epoxy groups. The organic Cl content is high (above 4%). The general yield of the diepoxy compound XXIX does not exceed 65%.

We described a different method of synthesis for the same diepoxy compound [20] (Scheme 35). The glycidyl ether is prepared directly from the alcohol XXX using excess ECH and the equimolecular alcohol/NaOH ratio in presence of iso-

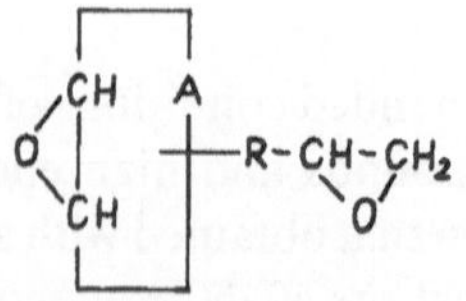

Scheme 33

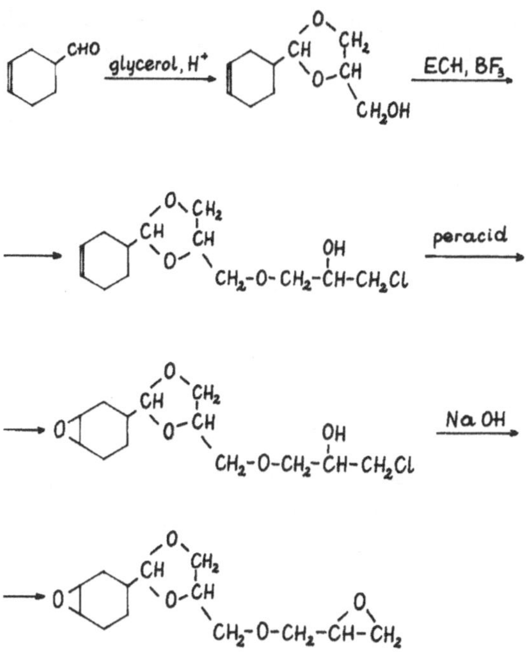

Scheme 34

XXIX

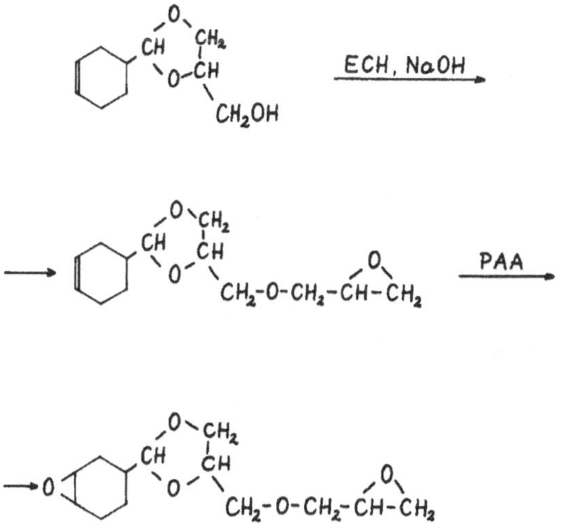

XXX

XXXI

XXXII

Scheme 35

propanol. The yield of the glycidyl ether XXXI is 93%, calculated on XXX, and the organic Cl content is low. Cyclohexene ring in the glycidyl ether XXXI is then epoxidized with aqueous PAA in benzene at 30–35 °C. No oxirane ring opening is observed. Yield of the diepoxy compound XXXII is 75%. The epoxy groups content amounts to 89% of the calculated value. Organic Cl content is a maximum of 0.3%.

Glycidyl derivatives of epoxycyclopentane were prepared in a similar way (Scheme 36)

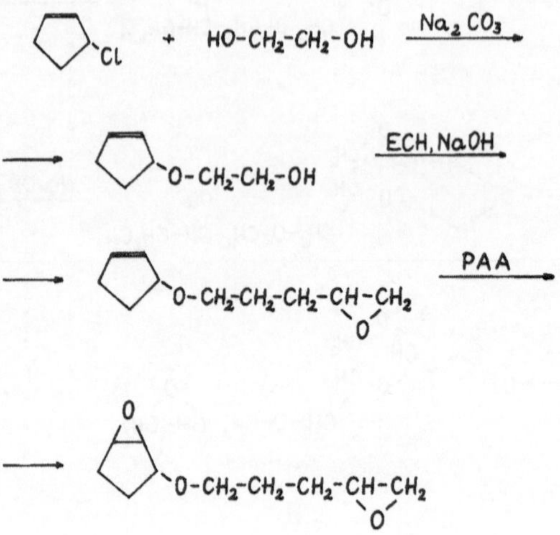

Scheme 36

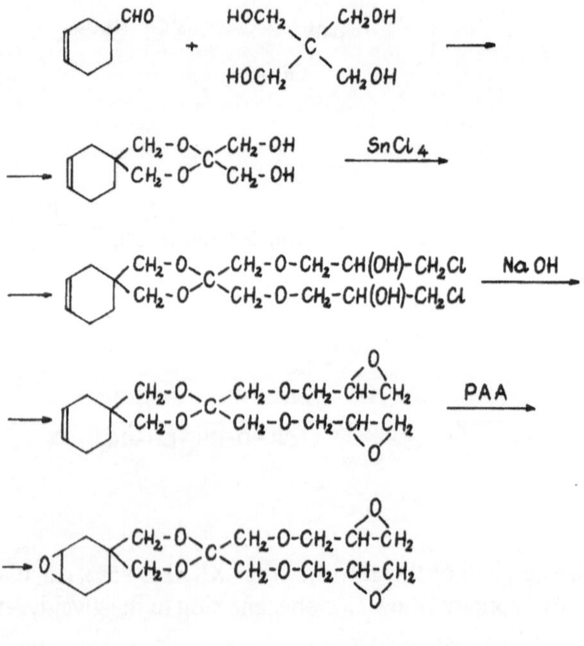

XXXIII

Scheme 37

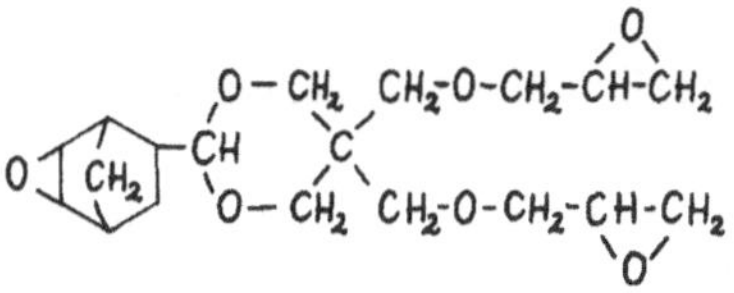

(XXXIV)

Scheme 38

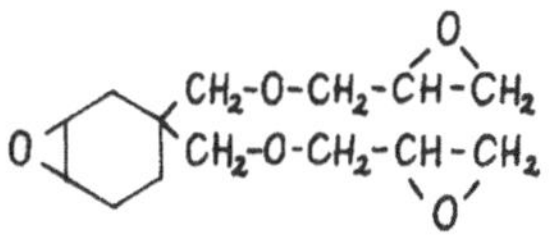

(XXXV)

Scheme 39

ACECs with three epoxy groups were synthesized starting from 3-cyclohexene-1-carboxaldehyde (1,2,3,6-tetrahydrobenzaldehyde) and pentaerythritol (Scheme 37) [4].The prepared monoacetal-diol was reacted with ECH in equimolecular ratio in the presence of SnCl4. The chlorohydrin ether was then treated with solid NaOH at 20–40 °C in an inert solvent (benzene, toluene). The diglycidyl ether obtained was eventually epoxidized with PAA at 30–35 °C. The triepoxide **XXXIII** was obtained.

Similar synthesis was performed using an endomethylene-cyclohexenyl derivative instead of the cyclohexenyl based triepoxide. The compound **XXXIV** (Scheme 38) was prepared.

We elaborated a simpler synthesis and isolation process for the triepoxides. The duration time of the process was shortened in comparison with that described in [19]. The yield of 1,1-bis(2,3-epoxypropoxymethyl)-3,4-epoxycyclohexane (UP-650 T) (**XXXV**, Scheme 39) was up to 85%.

Some properties of the ACEC triepoxides are given in Table 6..

Table 6. Aliphatic–cycloaliphatic triepoxides

Triepoxide	XXXIII	XXXIV	XXXV
Yield, %	80	78	85
Epoxy groups content, wt. %	26–32	25–29	36–39
Organic chlorine content, wt. %	2.0–3.0	2.5–3.5	1.0–2.0
Iodine number, mg I/g	1.0–1.8	2.0–3.0	–
Viscosity at 25 °C, Pas	5–9	7–10	0.2–0.3

3.3
Synthesis of Aliphatic-Cycloaliphatic Epoxy Compounds with Glycidyl Ester Groups

Synthesis of glycidyl esters of carboxylic acids is carried out by two main methods. The first one involves the reaction of ECH with the acid followed by dehydrochlorination of the chlorohydrin ester. The other method is based on the reaction of ECH with an alkali metal carboxylate. Other methods have only a preparative character.

The reaction of 3-cyclohexene-1-carboxylic acid with ECH (molar ratio 1:8) was investigated at 65–85 °C in presence of a catalyst (KCl or tetramethylammonium chloride). After 2 h the formation of the chlorohydrin ester was completed with 100% quantitative yield. However, the hydrochlorination results in a low yield of epoxy groups and a low conversion of chlorohydrin to epoxy groups. The alkali partial consumption for hydrolysis of the ester groups, even at room temperature, is another reason.

High yield (80–90%) of glycidyl 3-cyclohexene-1-carboxylate (**XVII**, Table 4) having a low organic chlorine content (below 1%) was achieved when the Na salt of the acid was reacted with ECH (molar ratio 1:14) in the presence of the above-mentioned catalysts. The epoxy groups content in the product amounted to 97% of the calculated value. A disadvantage of this method is that the Na salt has to be anhydrous.

Epoxidation of the glycidyl ester **XVII** thus prepared delivered glycidyl 3,4-epoxycyclohexane-1-carboxylate (**II**, Scheme 2) in 90% yield. The diepoxy compound had the following composition: epoxy groups content 40.0% (calculated 43.3%), organic Cl 0.37%, viscosity (at 25 °C) 26 mPa·s.

Epoxidation of diglycidyl esters **XXXVI** (Scheme 40) which can be synthesized directly from the corresponding dicarboxylic acids has been an object of considerable interest [21–23]. Such diepoxy compounds have found applications thanks to their low viscosity, high adhesion and good dielectric properties. However, there is a disadvantage in low heat resistance (TM below 100 °C) and unsatisfactory mechanical strength.

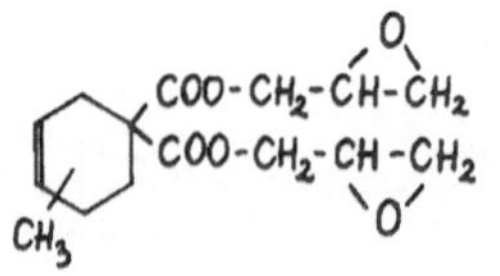

(XXXVI)

Scheme 40

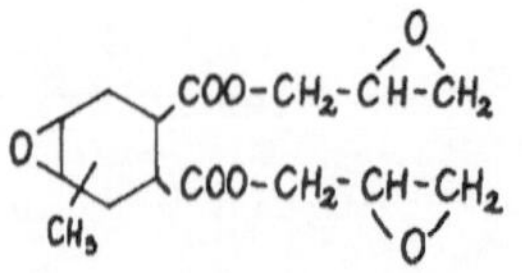

(XXXVII)

Scheme 41

Table 7. Diglycidyl 4,5–epoxycyclohexane–[3(4)–methyl]–1,2–dicarboxylates; XXXVII, Scheme 41

R	Formula	Oxirane oxygen content (%)		Organic Cl content (%)	Viscosity at 25 °C (Pa · s)	Yield %
		Calcd.	Found			
–	$C_{14}H_{18}O_7$	16.11	13.8–14.2	1.5–2.0	5.0–7.0	88
3–CH$_3$	$C_{15}H_{20}O_7$	15.38	13.4–13.8	1.5–2.0	4.5–7.0	83
3(4)–CH$_3$	$C_{15}H_{20}C_7$	15.38	12.5–13.4	1.5–2.0	4.0–6.5	82

In order to prepare products with improved properties, we tried to epoxidize first the double bond in diglycidyl 4-cyclohexene-1,2-dicarboxylates. Attempts to epoxidize diglycidyl 4-methyl-4-cyclohexene-1,2-dicarboxylate failed, presumably because of steric hindrance by the methyl group at the double bond. On the other hand, the epoxidation occurs easily in the case of the -CH=C(CH$_3$)-groups in an aliphatic chain [24].

There were no difficulties when 3-methyl substituted 4-cyclohexene-1,2-dicarboxylic acid or its diglycidyl ester was epoxidized with PAA. The yield and properties of the synthesized triepoxides (**XXXVII**, Scheme 41) are given in Table 7 [25, 26].

An improvement of the properties, achieved as a result of the increase in functionality of the simple ACECs, enabled a detailed study of the synthesis of more complicated ACECs with the functionality of at least 3. Easily available cycloaliphatic carboxylic acids were used as the starting materials.

A series of triepoxy ACECs of the general formula **XXXVIII** (Scheme 42) was synthesized [27]. In the general formula **XXXVIII**, R is H or CH$_3$, X is -CH$_2$-, -CH$_2$CH$_2$-O- or the dioxolane bridge **XXXIX** (Scheme 43) and Y is epoxycyclopentyl or epoxycyclohexyl. An example of such a compound (**XL**) is shown

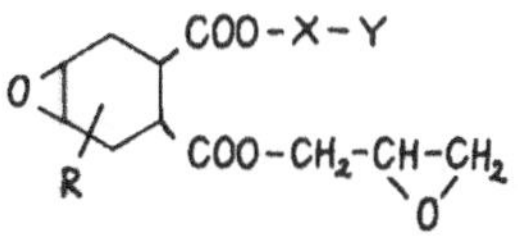

(XXXVIII)

Scheme 42

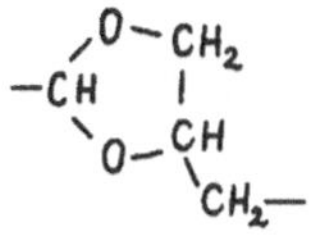

(XXXIX)

Scheme 43

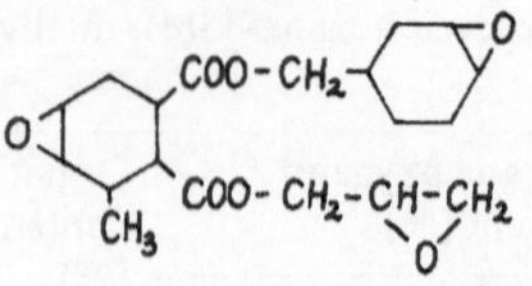

(XL)

Scheme 44

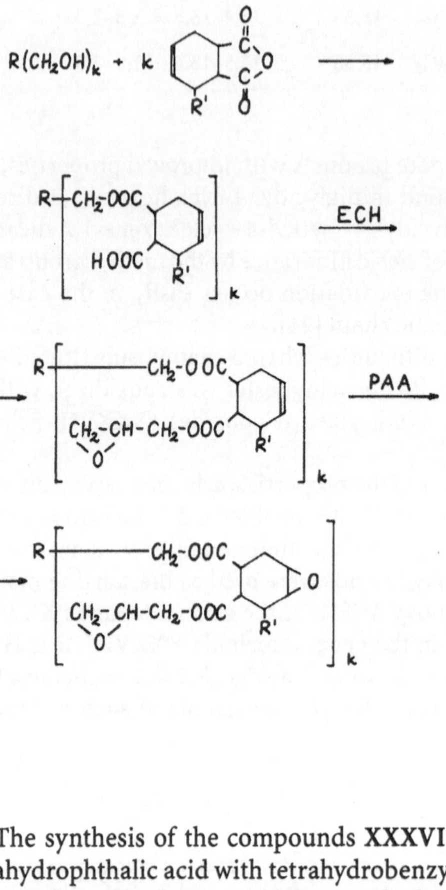

Scheme 45

above (Scheme 44). The synthesis of the compounds **XXXVIII** starts from the condensation of tetrahydrophthalic acid with tetrahydrobenzyl alcohol, tetrahydrobenzalglycerol or ethylene glycol monocyclopentenyl ether.

Another approach to the synthesis of higher functionality ACECs is shown in Scheme 45; $R(CH_2OH)_k$ is polyhydric alcohol (a glycol, trimetylolpropane or pentaerythritol); k is 2, 3 or 4; and R is H or CH_3 [27]. If $R(CH_2OH)_k$ is 1,1-di(hydroxymethyl)-3-cyclohexene, the epoxidation product contains one more epoxycyclohexane group.

The following reactions (Scheme 46) lead to another group of ACECs which contain epoxycyclohexane and glycidyl ester groups. The polyhydric alcohols $R(CH_2OH)_{k+m}$ and the group R' have the same meaning as given above; k and m are 1, 2 or 3 and k+m is 2, 3 or 4.

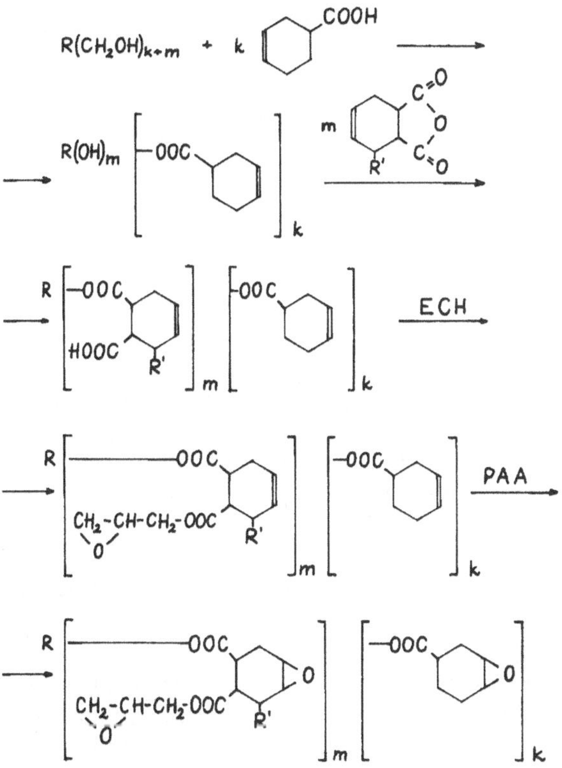

Scheme 46

The first stage of the synthesis consists of the addition of the acid anhydride to hydroxyl groups in the polyhydric alcohol or to the partial esterification product. The addition is carried out at 145–175 °C. The synthesis of the glycidyl esters involves a reaction of the formed COOH groups with excess ECH in presence of catalytic amounts of KCl and water, up to the complete consumption of the carboxyl groups. The dehydrochlorination is carried out with solid NaOH at 40–50 °C. The epoxidation of cyclohexene rings with PAA is the final step of the synthesis. The yield of the whole process is 77–86%. The products are highly viscous liquids with an oxirane groups content (both glycidyl and epoxycyclohexane) of 9.3–12.7%.

The above schemes of ACECs synthesis make it possible to control easily the distance between epoxycyclohexane rings and the structure of the linkages between the rings. Moreover, the ratio of epoxy groups in the aliphatic (i.e. glycidyl) and the cycloaliphatic (i.e. epoxycyclohexane) part of the molecule can be changed. The presence of epoxy groups differing in structure and reactivity with curing agents makes possible a controlled crosslinking involoving the stepwise entering of different epoxy groups into reaction with the curing agent.

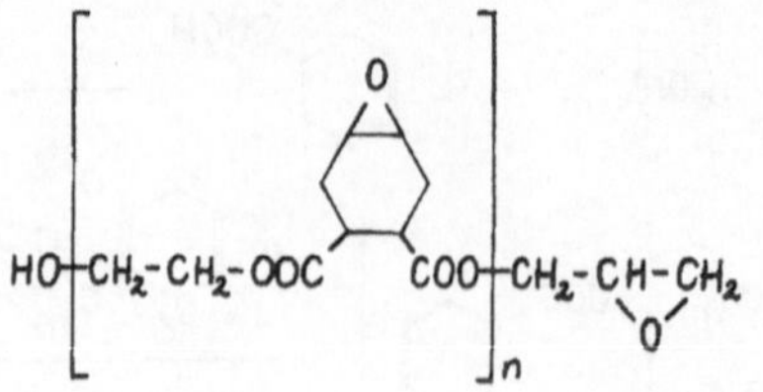

Scheme 47

Similar ACECs (Scheme 47) were prepared in a different way [28]. The carboxylic polyesters were prepared by polycondensation of 4-cyclohexene-1,2-dicarboxylic acid anhydride with a glycol. Then the glycidylation of COOH groups with ECH followed by epoxidation of cyclohexene rings with PAA was carried out. Depending on the COOH groups content in the intermediate polyester, various glycidyl/epoxycyclohexane ratios can be obtained.

3.4
Synthesis of Aliphatic-Cycloaliphatic Epoxy Compounds with Glycidyloxyphenyl Substituents

It was assumed that aromatic rings in the ACECs should improve the heat resistance of the crosslinked polymers. A series of ACECs containing epoxycyclohexane groups and glycidyl ethers of phenols in the molecule were prepared. One of the possible syntheses starts from a hydroxybenzaldehyde and 1,1-di(hydroxymethyl)-3-cyclohexene (Scheme 48) [29]. A similar product (Scheme 49) was prepared using an intermediate compound synthesized from pentaerythritol and 3-cyclohexene-1-carboxaldehyde.

Glycidyl ethers of the cycloacetal-phenols were prepared in the usual way by reaction with excess ECH followed by dehydrochlorination at 75–85 °C. The epoxy groups content in the glycidyl ethers and in the diepoxides (after the epoxidation with PAA) was above 90% of the calculated value.

Another process consists of building the epoxycyclohexane unit into the bridge of a bisphenol molecule (Scheme 50) [30]. The condensation of phenol with 3-cyclohexene-1-carboxaldehyde may occur in the presense of both acid and alkaline catalysts. If oxalic acid was used as the catalyst, the yield of the bisphenol **XLI** reached 60%. Still higher yield was reached when excess phenol in the presence of HCl as catalyst was applied. The bisphenol **XLI** (melting temp. 318.5 °C) was obtained by crystallization from isopropanol. The p,p'-isomer contained small amounts of the o,p'-and o,o'-isomer [31]. The structure of the product was confirmed by IR spectrophotometry and molecular weight determination.

Diglycidyl ether of bisphenol **XLI** was prepared by the reaction of bisphenol with excess ECH followed by dehydrochlorination with solid NaOH. The yield was 92–95% and the oxirane oxygen content was 7.50% (calculated 8.18%). Epoxidation of the diglycidyl ether **XLII** with PAA at 30–35 °C yielded the trie-

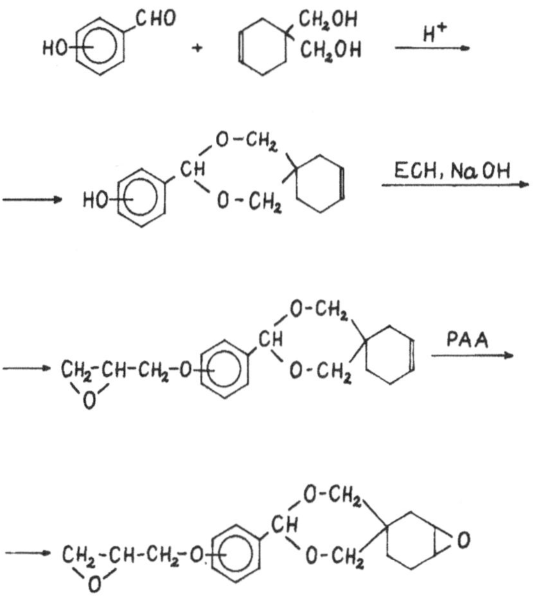

$$HO-C_6H_4-CHO + \text{(cyclohexenyl)}\underset{CH_2OH}{\overset{CH_2OH}{}} \xrightarrow{H^+}$$

$$\xrightarrow{} HO-C_6H_4-CH\underset{O-CH_2}{\overset{O-CH_2}{}}\text{(cyclohexenyl)} \xrightarrow{ECH,\ NaOH}$$

$$\xrightarrow{} CH_2-CH-CH_2-O-C_6H_4-CH\underset{O-CH_2}{\overset{O-CH_2}{}}\text{(cyclohexenyl)} \xrightarrow{PAA}$$
with epoxide on CH₂-CH-CH₂ (O bridge)

$$\xrightarrow{} CH_2-CH-CH_2-O-C_6H_4-CH\underset{O-CH_2}{\overset{O-CH_2}{}}\text{(epoxycyclohexyl)}$$

Scheme 48

$$CH_2-CH-CH_2-O-C_6H_4-CH\underset{O-CH_2}{\overset{O-CH_2}{}}C\underset{CH_2-O}{\overset{CH_2-O}{}}\text{(epoxycyclohexyl)}$$

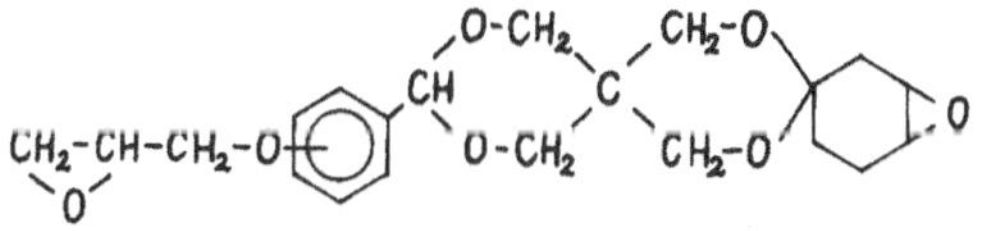

Scheme 49

poxide **XLIII**, softening temperature 55–60 °C, oxirane oxygen content 11.20% (calculated 11.79%); yield 85–90%.

The occurrence of epoxidation was confirmed by IR spectrophotometry: disappearance of double bonds in the cycloaliphatic ring (1665 and 3040 cm–1) and appearance of the absorption at 905 cm^{-1} (oxirane at the six-membered ring).

There is no noticeable acidolysis of the epoxy groups in the course of the epoxidation: no absorption at 1720 cm^{-1}, which could be attributed to ester groups, was observed.

An additional method of ACEC synthesis was based on hydroxybenzoic acids as the starting materials [32]. The synthesis starts from esterification of the acid with 3-cyclohexenyl-1-methyl alcohol. The following hydroxybenzoic acids were used: 2-hydroxybenzoic (salicylic), 2,4-di-hydroxybenzoic (β-resorcylic) and 3,4,5-trihydroxybenzoic (gallic) acid. The phenolic hydroxyls in the cyclohexenylmethyl esters were transformed in glycidyl ethers in the usual way (reaction with excess ECH followed by dehydrochlorination). Eventually, epoxidation with PAA was carried out (Scheme 51). The epoxide groups content in the ACECs thus prepared was above 85% of the calculated value.

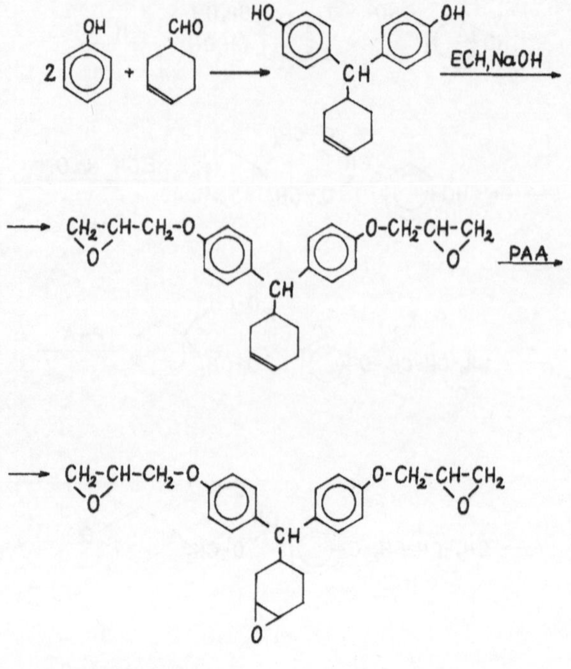

Scheme 50

(XLII)

(XLIII)

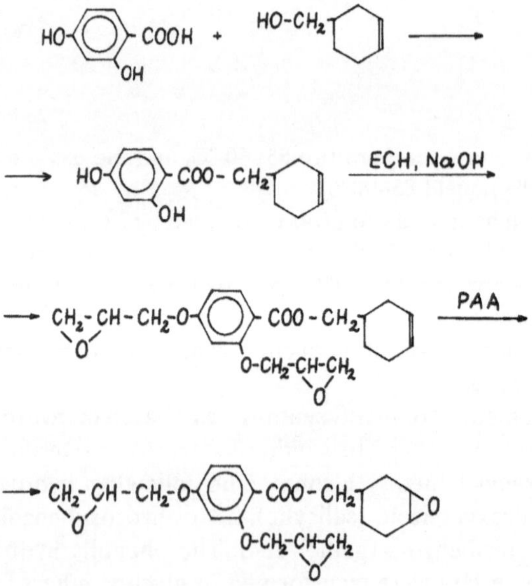

Scheme 51

3.5
Imide Rings Containing Cycloaliphatic EpoxyCompounds

An increase in heat resistance, thermal stability and mechanical strength of ACECs by introducing aromatic rings has been desirable. It can be assumed that cyclic systems involving imide rings condensed with aromatic or epoxycycloalkane rings meet the above equirements. A series of imide rings containing ACECs was synthesized (Scheme 52) [33]. The condensation (imidization) step was carried out in boiling xylene. Water was distilled off as an azeotrope. The imidization was completed at 180 °C. The glycidyl ethers were prepared by the conventional method using excess ECH (molar ratio ECH:phenolic OH=10:1) and with the stepwise addition of aqueous NaOH solution. Then the epoxidation of the double bond with PAA was carried out. The oxirane oxygen content was 7.50% (calculated 10.70%), organic Cl 1.2%.

A large group of CECs with imide rings was synthesized by the addition of an unsaturated imide and an imide ring containing unsaturated glycidyl ether followed by the epoxidation of the unsaturation in cyclohexenyl rings [34, 35] (Scheme 53). The imide/glycidyl ether addition was carried out at 140 °C/2–3 h in the presence of tertiary amines as the catalysts. 3-Methyl substituted cyclohexenyl derivatives were also used.

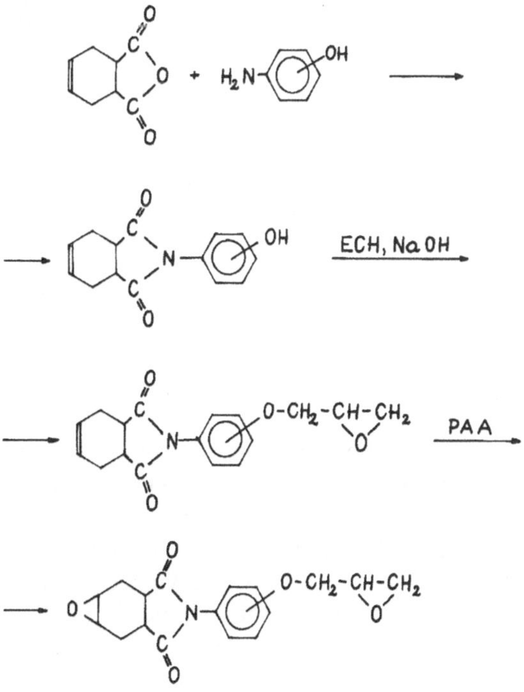

Scheme 52

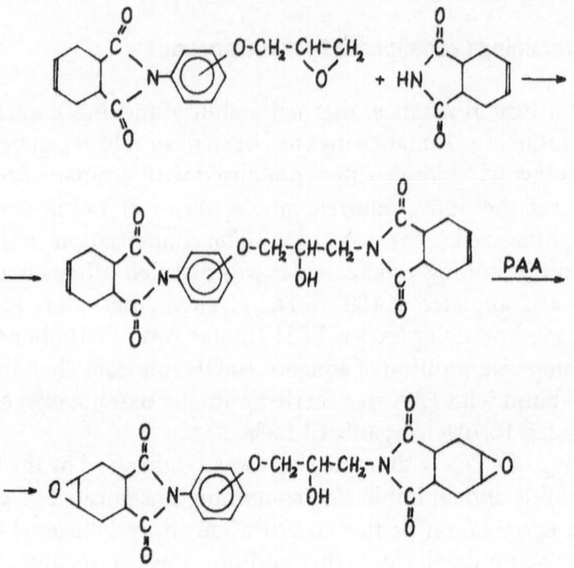

Scheme 53

The prepared CECs are highly viscous liquids or solids with a softening temper-ature below 60 °C. The epoxy groups content is above 90% of the calculated value. The IR spectrophotometry confirms the occurence of imide rings (1785, 1720 and 1420–1375 cm^{-1}) and epoxy groups at cyclohexane ring (790–795 cm^{-1}).

The imide rings containing CECs described above do not involve glycidyl groups. However, they contain 2-hydroxy-1,3-propylene bridges which are formed from glycidyl groups and can thus be considered to be related to ACECs. The same concerns the phosphorus-containing CECs which are presented be-low.

3.6
Phosphorus-Containing Cycloaliphatic Epoxy Compounds

There is little information about phosphorus containing CECs in the litera-ture.The synthesis of phosphoric CECs is based on the addition of phosphite es-ters and glycidyl ethers or esters. Both reactants contain cyclohexene rings which are further epoxidized with PAA in CHCl$_3$ (Scheme 54).

The acid phosphite is easily prepared in high yield from equimolar amount of the diol, PCl$_3$ and ethanol. Strong absorption bands in the IR spectra appear at 1035 cm^{-1} (P-O-C), 1285 cm^{-1} (P→O), 2405 cm^{-1} (P-H). Medium intensity ab-sorption bands appear at 600 and 3030 cm^{-1} (double bond in the cyclohexene ring).

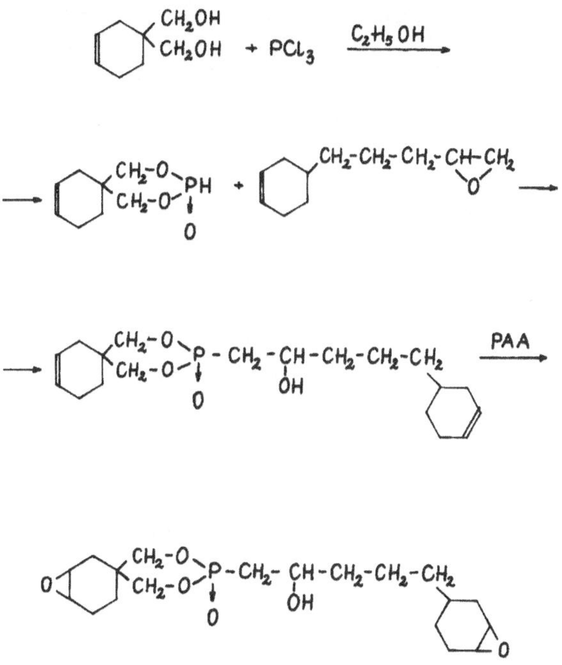

Scheme 54

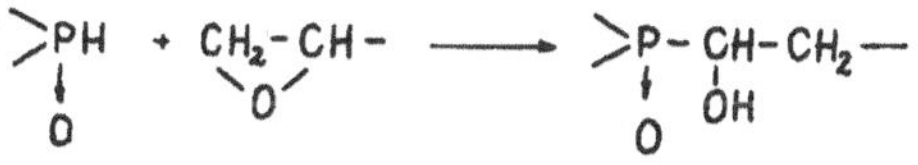

Scheme 55

The addition of P-H and the glycidyl group occurs only in presence of Na metal as a catalyst [36]. *N,N*-Dimethylbenzylamine (0.5–1.0%, 120–140 °C) is also effective as the catalyst. The disapperance of epoxy groups corresponds quantitatively to the formation of secondary OH groups (Scheme 55).

The above reactions (Schemes 54 and 55) lead first to the formation of the diene **XLIV** followed by the epoxidation with PAA in chloroform. The diepoxide **XLV** is obtained in around 80% yield.

3.7
Aliphatic-Cycloaliphatic Epoxy Oligomers

Flexible epoxy resins were synthesized by the polymerization of epoxy compounds in the presence of an unsaturated alcohol (Scheme 56) followed by the epoxidation of the double bond with PAA. If a glycidyl ether of a cyclohexenyl ring-containing alcohol is used as the epoxy compound, a polyether-epoxy resin is obtained (Scheme 57). The polymerization of 3-cyclohexenyl-1-methyl glyci-

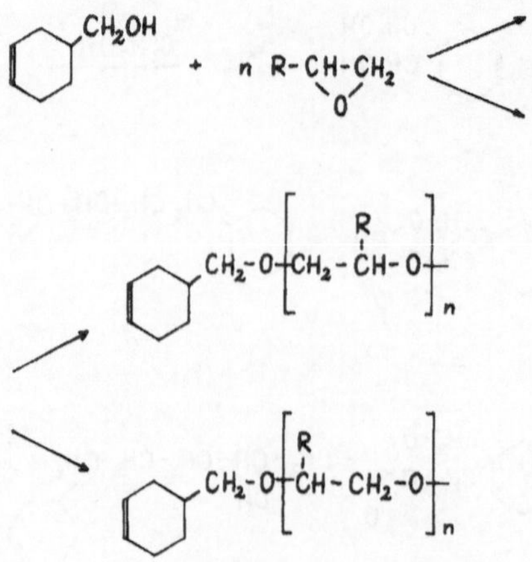

Scheme 56

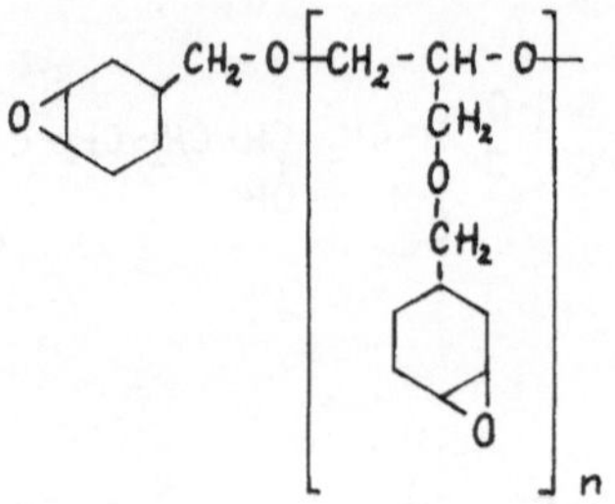

Scheme 57

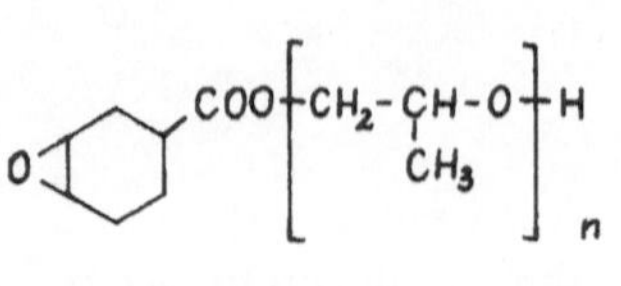

Scheme 58

dyl ether was carried out in an organic solvent at 80 °C in the presence of 2–5% water and an acid activated clay. After the epoxidation with PAA, low viscous oligomers with the oxirane oxygen content 6.0–8.2% and OH content 2.5–4.5% are formed. The polymerization degree depends on the ratio of the reactants and can be easily controlled [37].

The polyether-epoxy flexibilizers which contain epoxy groups along the chain (Scheme 57) impair better mechanical properties of the crosslinked composition in comparison with mixtures of aliphatic epoxy resins and polyether polyols. The improved properties are also obtained using 3,4-epoxycyclohexane-1-carboxylates of polyoxypropylene polyether polyols [38] (Scheme 58).

If the glycidyl ether used for the synthesis of the polyether-epoxy resins (Scheme 57) is replaced by a similar glycidyl ester (glycidyl 3-cyclohexene-1-carboxylate), the polymerization reaction at 100 °C becomes out of control. This

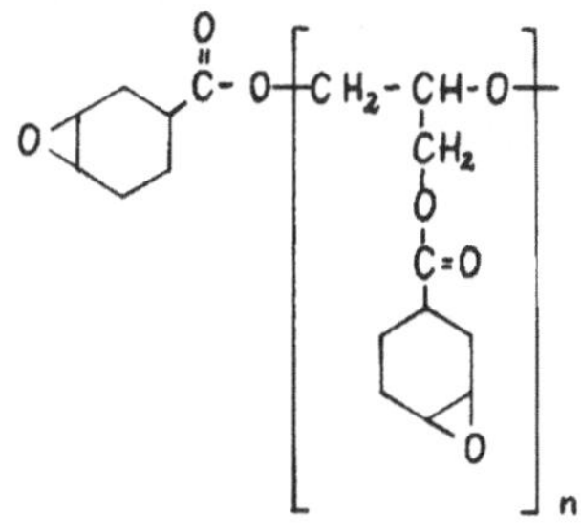

Scheme 59

phenomenon is caused by a strong electron acceptor effect of the -COO- group which results in an increase in basicity of the epoxy group. However, in the presence of tertiary amines as catalysts a normal course of polymerization at 90–100 °C is observed. The cyclohexenyl units in the oligomers which do not contain glycidyl groups can be epoxidized with PAA (Scheme 59).

The relatively low yield (60%) in the epoxidation stage is due to the losses of the highly polar oligomers as a result of their solubility in the aqueous phase.

3.8
Brominated Analogues of Cycloaliphatic Epoxy Compounds

There is very little information about halogenated aliphatic and cycloaliphatic epoxy compounds. Brominated epoxy compounds were prepared by the addition of Br to the cyclohexenyl double bond in a diglycidyl ether in mild conditions (Scheme 60) [39].

The synthesized dibromodiepoxide contains 38.10% Br (calculated 38.65%) and 7.26% oxirane oxygen (calculated 7.73%) and is a liquid with a medium viscosity (2.23 Pa at 25 °C), $d_4^{20} = 1.552$ and $n_D^{20} = 1.5320$ [40].

Bromination of diglycidyl ester of 4-hexene-1,2-dicarboxylic acid under similar conditions (addition of stoichiometric amount of bromine to the solution of

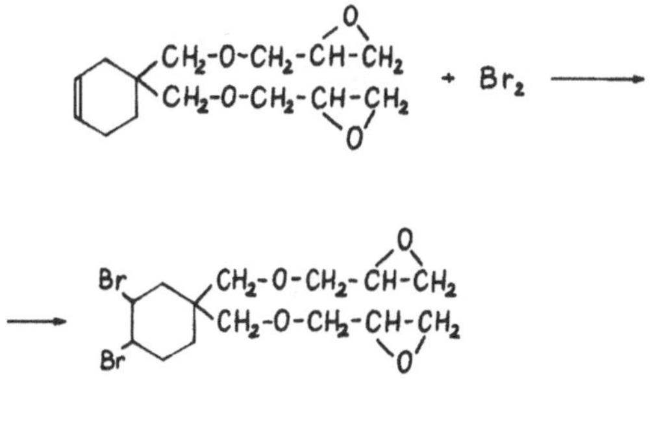

Scheme 60

the unsaturated diepoxide in $CHCl_3$ or CCl_4 at 10–40 °C) results in parallel formation of bromohydrins by the competitive reaction of substitution bromination in the glycidyl epoxy groups.

3.9
Nitrogen Containing Aliphatic-Cycloaliphatic Epoxy Compounds

Synthesis of cyanoethylated ACEC **XLVI** (Scheme 61) involves the preparation of the monoglycidyl ether of the diol followed by cyanoethylation with acrylonitrile in the presence of 40% NaOH as catalyst (87% yield) and epoxidation with 45% aqueous PAA (89% yield). The diepoxide contains 9.93% oxirane oxygen (calculated 12.0%).

Properties of a composition consisting of a standard bisphenol A/ECH epoxy resin with 20% cyanoethylated ACEC (**XLVI**) and an acid anhydride curing agent exhibit a considerably elevated heat resistance.

The different reactivity of the aliphatic and cycloaliphatic epoxy groups in ACECs can be used for the synthesis of functionalized compounds with the cycloaliphatic epoxy groups preserved. As an example, an epoxy group containing polyhydric alcohol **XLVII** is presented (Scheme 62). The epoxy polyol **XLVII** was applied for the modification of isocyanates on purpose to obtain flexible coatings with good protective properties.

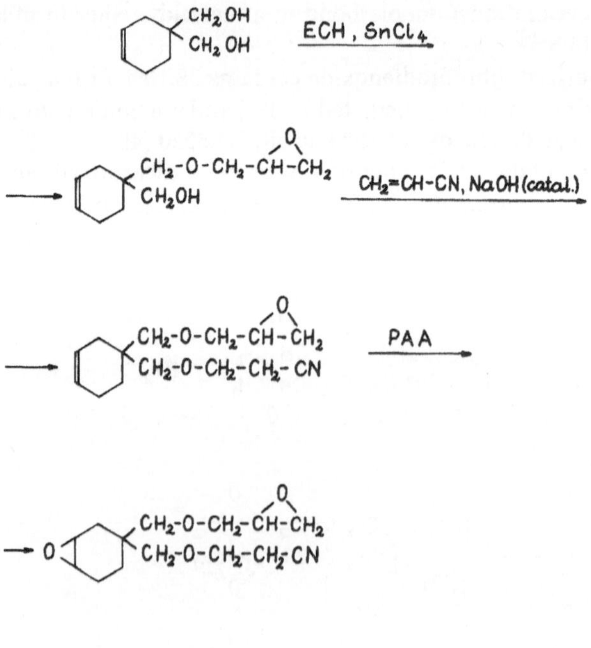

(XLVI)

Scheme 61

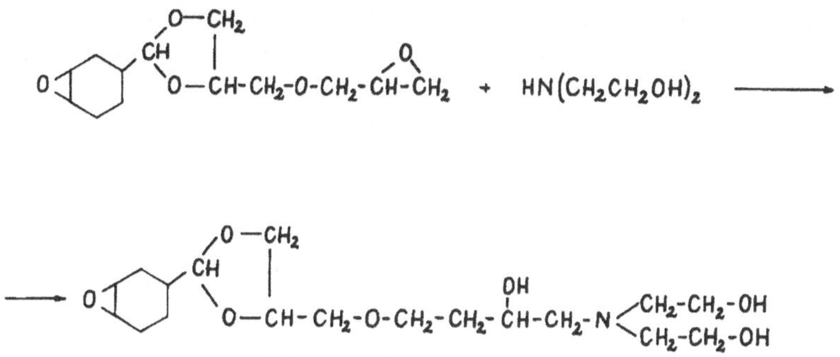

(XLVII)

Scheme 62

4
Effect of the Chemical Structure of Aliphatic-Cycloaliphatic Epoxy Compounds on the Properties of the Crosslinked Polymers

4.1
Reactions of Aliphatic-Cycloaliphatic Epoxy Compounds with Acidic and Basic Reagents

Various methods for synthesis of ACECs make it possible to create a broad assortment of novel products to be applied as binders in polymer composites.

The correlation between the chemical structure of the components, in particular of systems consisting of ACECs, curing agents and accelerators, and the properties of polymeric materials has a multifactorial character [41]. The following factors are involved:
- the cycloaliphatic/glycidyl epoxy groups ratio,
- chemical structure of the linkages between the glycidyl units and the remaining fragments of the ACEC molecule,
- chemical structure of the fragment of the ACEC molecule between the reactive units,
- kinetic characteristics of the reaction of different classes of curing agents with ACECs.

The kinetic investigations were correlated with the above-mentioned structural features.

The other factors which affect the properties of crosslinked polymers such as the molar ratio of the functional groups, the curing time and temperature etc. were kept constant for the individual classes of the curing agents.

It has been shown that the extension of the chain between the epoxycyclohexane rings in CECs and the incorporation of aliphatic chains in the molecule of the curing agent result in a decrease in the heat resistance [41–44]. However, the mechanical properties are only a little improved. The reactivities of individual epoxy groups in CECs differ insignificantly. The crosslinking regularity [45] and the properties of the crosslinked polymeric materials cannot thus be affected by an improvement of the chemical structure of CECs [46], even if flexibilizers are used [47].

Quantitative characteristics of the reactivity of epoxycyclohexane and glycidyl groups was studied using model compounds [48–50]. The effect of nucleophilic substituents was studied, taking into account the influence of the nucleophilic substitution on the basicity of the epoxy ring.

It was important to find to what extent the relationships concern the real ACECs being cured with standard hardeners in the presence of accelerators, e.g. tertiary amines and tertiary amine substituted phenols. To start with, the addition of ACECs (Table 8) with model hardeners 3-cyclohexene-1-carboxylic acid and aniline was studied.

The commercial ACECs investigated can be considered as individual chemical compounds. The epoxy groups (in the ACEC), carboxyl group (in the acid) or active hydrogen atoms (in aniline) ratios were stoichiometric. The reaction with the acid and with aniline was carried out at 80 and 100 °C, respectively. The addition of the acid was accelerated with N,N-dibenzylamine (1% of the epoxy component).

Typical kinetic curves exemplifying the amine catalyzed reactions of ACECs with the carboxylic acid are presented in Fig. 7. The reaction rate of the cycloaliphatic epoxy groups in ACECs is higher than that of the glycidyl groups. The effect of ether and ester bonds on the reactivity of these epoxy groups is different. The ester bond, which is a strong electron acceptor, diminishes the basicity of cycloaliphatic epoxy groups. This results in a decrease in reactivity of the epoxy groups in relation to carboxylic acids in both non-catalyzed [51] and catalyzed additions. On the other hand, the reactivity of the epoxy groups in glycidyl esters is higher than that of glycidyl ethers. This is why the differences between the reactivity of cycloaliphatic and glycidyl epoxy groups with the carboxylic acid are more distinct in the case of the glycidyl ethers UP-656 and UP-650 T in comparison with the corresponding glycidyl esters UP-656 M and UP-640 T.

Table 8. Properties of some commercial ACECs (see also Tables 9 and 10)

Com- pound	Scheme	Commercial product	Formula	Molecular weight		Oxirane oxygen content (%)	
				Calcd.	Found	Calcd.	Found
I	1	UP–656	$C_{10}H_{16}O_3$	184	190	17.4	16.8
II	2	UP–656M	$C_{10}H_{14}O_4$	198	205	16.2	15.9
XXXVII[a]	41	UP–650T	$C_{14}H_{22}O_5$	270	295	17.8	16.5
XXXVII[b]	41	UP–64OT	$C_{15}H_{20}O_7$	312	337	15.4	14.1

[a]Without methyl group
[b]With 4–methyl group

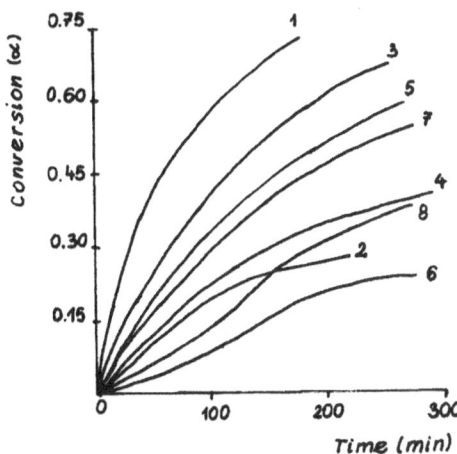

Fig. 7. Conversion (α) of the cycloaliphatic (CA) and glycidyl (GL) epoxy groups in course of the addition of 3-cyclohexene-1-carboxylic acid at 80 °C: 1 – 656, CA; 2 – 656, GL; 3 – 656 M, CA; 4 – 656 M, GL; 5 – 650 T, CA; 6 – 650 T, GL; 7 – 640 T, CA; 8 – 640, GL. The numbers "656" etc. relate to the commercial ACECs "UP" given in Table 8

As far as the addition of aniline to the ACECs is concerned, the principal conclusions in [50] were confirmed: glycidyl type epoxy groups are more reactive than those attached to the cyclohexane ring. Moreover, there is a very distinct induction period in the case of glycidyl groups. The appearance of the induction period is connected with the accumulation of hydroxyl groups which serve as an accelerator of the epoxy-amine addition. This acceleration effect is little noticeable as far as cycloaliphatic epoxy groups are concerned.

4.2
Crosslinking of Aliphatic-Cycloaliphatic Epoxy Compounds with Acid Anhydrides and Amines

It was confirmed that it was possible to consume different epoxy groups consecutively in the course of curing [52, 53]. In the early stage of curing, reaction of carboxyl groups with cycloaliphatic epoxy groups prevails, resulting in the formation of a polymer chain with a loose crosslinking. In the later stages, the chain extension and dense crosslinking proceeds as a result of the conversion of glycidyl groups and of the remained cycloaliphatic epoxy groups. Eventually, the network density is achieved. The network density is determined by the ACEC-hardener ratio and by the conditions of the curing process.

The reaction sequence is different if amines are applied as curing agents. In the first stage the glycidyl groups react. Then the cycloaliphatic epoxy groups enter into the reaction with the curing agents.

Table 9. Properties of amine cured ACECs (see Table 7)

Compound (UP-)	Chemical structure				TM (°C)	Mechanical strength (MPa)			Elongation at break (%)
	Epoxy groups					Tensile	Flexural	Compression	
	Glycidyl ether	Glycidyl ester	Cycloaliphatic	Glycidyl					
656	1	–	1	1	80//74	53//58	94//75	161//120	1.8/1.9
656M	–	1	1	1	90//80	55//53	106//80	163//157	2.5//2.1
650T	2	–	1	2	10//98	81//92	124//93	172//154	3.4//3.4
640	–	2	1	2	134//104	86//100	131//12	174//164	3.6//3.2

Curing with 4,4'-diaminodiphenylmethane // with 3,3'-dichloro-4,4'-diaminodiphenylmethane

Nevertheless, the sequential entering of different epoxy groups into the reaction, irrespective of the acidic or basic character of the curing agent, is a very important feature of the crosslinking process of ACECs because it conditions the formation of a regular polymer network [41].

High conversion of cycloaliphatic epoxy groups in the triepoxide ACECs on curing with aromatic amines is noticeable. This phenomenon can be attributed to the catalytic effect of secondary hydroxyl groups being formed as a result of the glycidyl-amine addition as well as – to a lesser extent – the increase in reactivity as discussed above. All this secures an increased mechanical strength and heat resistance of cured ACECs (Table 9).

The crosslinked triepoxides exhibit higher mechanical strength and higher heat resistance in comparison with the diepoxy ACECs. The heat resistance of acid anhydride cured triepoxides is particularly high. This feature is obviously caused by the higher crosslinking density.

Mechanical properties of crosslinked diepoxide with two equal glycidyl groups (the diglycidyl ether UP-650 and the diglycidyl ester UP-640) exhibit higher mechanical strength values than the diepoxides with one cycloaliphatic epoxy group and one glycidyl group (the monoglycidyl ether UP-656 and the monoglycidyl ester UP-656 M).

No distinct correlation could be found between the relative functionalities of $6.45 \cdot 10^{-3}, 9.0 \cdot 10^{-3}, 6.9 \cdot 10^{-3}$ and $10.0 \cdot 10^{-3}$ of the commercial products UP-640, UP-640 T, UP-650 and UP-650 T calculated from the molecular weight values [54] and the mechanical strength and heat resistance of the crosslinked polymers (Table 10). It should be mentioned that the building of epoxy groups into the cyclohexane ring results in a considerable increase in the heat resistance, good mechanical strength being preserved. The character of the linkage between the glycidyl group and the rest of the molecule has a considerable effect on properties of the crosslinked epoxy compounds.

The comparison of the properties of amine cured diepoxy compounds containing one epoxycyclohexane and two glycidyl ether group (UP-650 T) with the properties of a diglycidyl ether (UP-650) shows that the appearance of one epoxy group at a cyclohexane ring contributes to higher mechanical strength and heat resistance, independently from the chemical structure of the amine curing agent [55] (Table 11). This feature can be attributed to the above-mentioned consecutive entering of various epoxy groups into the reaction with the curing agent which may result in a certain structural regularity of the polymer network.

The heat resistance values of the aromatic amine cured ACECs was, however, lower than expected from their relative functionality. Incomplete conversion of the cycloaliphatic epoxide groups on curing with aromatic amines was assumed to be the reason. Therefore, an addition of an acidic curing agent, e.g. a bisphenol, was expected to result in an increase in conversion of the epoxycyclohexane groups. The addition of resorcinol to the compositions consisting of the epoxy compounds XLVIII and IL (Schemes 63 and 64) (one cycloaliphatic epoxy group and two glycidyl ester groups) indeed has a favourable influence on the mechan-

Table 10. Properties of acid anhydride cured ACECs and related epoxy compounds (see Table 7)

| Compound (UP-)[a] | Chemical structure | | | Acid anhydride curing agent[b] | TM (°C) | Mechanical strength (MPa) | | | Elongation at break(%) |
	ether	ester	cycloaliphatic epoxy groups			Tensile	Flexural	Compression	
650T XXV	2	–	1	A	162	58	100	148	2.1
				B	142	79	79	150	3.8
				C	–	20	22	50	24.0
650 LIX	2	–	–	A	120	60	110	138	3.2
				B	116	58	105	142	2.8
				C	–	20	20	40	23.0
656 I	1	–	1	A	96	44	80	120	1.5
				B	98	52	80	128	2.2
				C	–	21	20	60	28.0
640 XIX	–	2	–	A	93	81	99	124	3.8
640T XXXVII	–	2	1	A	172	82	94	157	2.2

[a]Epoxy compounds:
UP-650T: diglycidyl ether of 1,1–di(hydroxymethyl)–3,4–epoxycyclohexane
UP-650: diglycidyl ether of 1,1–di(hydroxymethyl)–3–cyclohexene
UP-656: glycidyl ether of 3,4–epoxy–1–hydroxymethyl–cyclohexane
UP-640: diglycidyl ester of 3(4)–methyl–4–cyclohexene–1,2–dicarboxylic acid
UP-640T: diglycidyl ester of 4,5–epoxy–3(4)–methyl–cyclohexane–
   1,2–dicarboxylic acid

[b]Acid anhydride curing agents:
A : 4–methyl–4–cyclohexene–l,2–dicarboxylic acid anhydride
B : methyl–endomethylene–tetrahydrophthalic anhydride
   (Methyl Nadic Anhydride)
C : polyadipic polyanhydride

Table 11. Effect of the amine curing agents on the properties of crosslinked epoxy compounds

Curing agent	Epoxy compound (UP–)	TM (°C)	Mechanical strength (MPa)			Elongation at break %
			Tensile	Flexural	Compression	
MOCA	650	95	95	95	156	1.7
	650T	78	65	68	146	3.5
DDS	650	155	60	74	172	2.7
	650T	120	58	60	148	2.0
DDM	650	140	90	100	210	3.0
	650T	115	67	82	160	3.0
TEAT	650	90	73	111	210	3.0
	650T	80	60	85	170	4.5

Amine curing agents:
MOCA: 3,3'-dichloro-4,4'-diaminodiphenylmethane
DDS: 4,4'-diaminodiphenylsulfone
DDM: 4,4'-diaminodiphenylmethane
TEAT: triethanolamine titanate
Epoxy compounds: see Table 10

ical strength values and elongation at break of the crosslinked polymer (Table 12) [25].

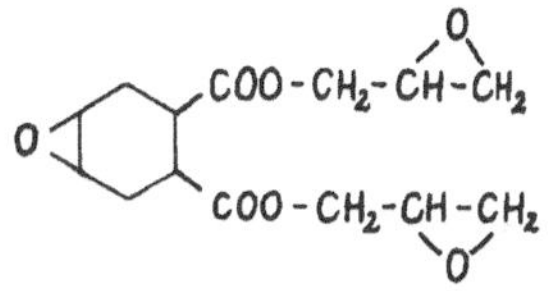

(XLVIII)

Scheme 63

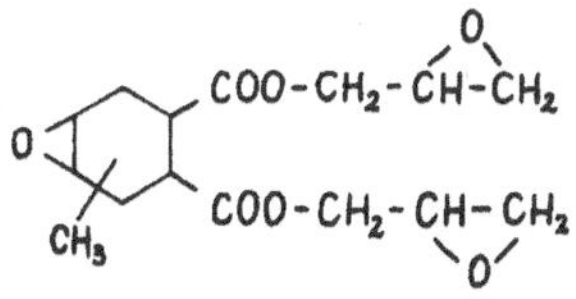

(IL)

Scheme 64

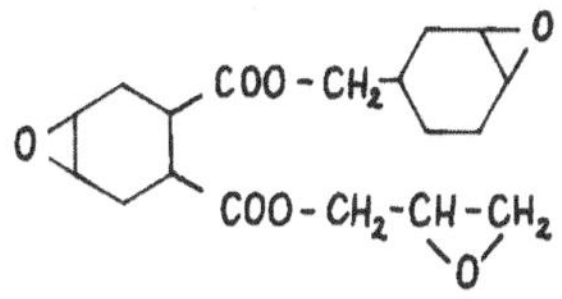

(L)

Scheme 65

Table 12. Properties of cured diglycidyl esters

Epoxy compound (Schemes 63 and 64)	Curing agent						Mechanical strength (MPa)			Elongation at break (%)	Water absorption 24 h/25 °C
	m-Phenylene diamine	Resorcinol	EUT	MOCA	3,3'-DDS	4,4'-DDS	Tensile	Flexural	Compression		
XLVIII	+						82	170	152	9.4	0.32
IL		+		+			65	130	168	3.8	–
IL		+	+				61	128	171	1.9	–
IL			+	+			88	123	178	3.8	–
IL		+		+			82	122	167	3.0	–
IL		+		+			110	190	195	5.1	0.09
XLVIII		+	+	+			88	156	201	–	–
IL		+	+	+			106	171	196	–	–
IL		+			+		91	150	210	–	–
IL		+				+	90	115	203	–	–

EUT: liquid eutectic mixture of aromatic amines
MOCA: 3,3'-dichloro-4,4'-diphenylmethane
DDS: diaminodiphenylsulfone

The investigation of crosslinked multifunctional ACECs (Table 13) has shown that the increase in the number of glycidyl or cycloaliphatic epoxy groups in an ACEC molecule does not necessarily improve the properties of the crosslinked polymer. The increase in functionality and the dense structure of the molecule between the functional groups results in a limited mobility of the reacting units and, consequently, in an incomplete conversion of the functional groups and

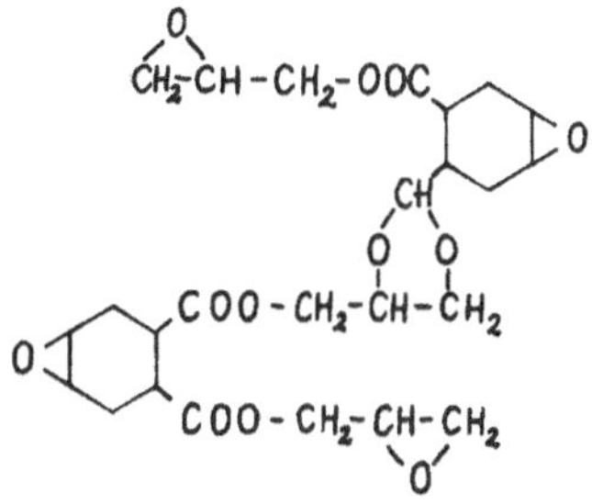

(LI)

Scheme 66

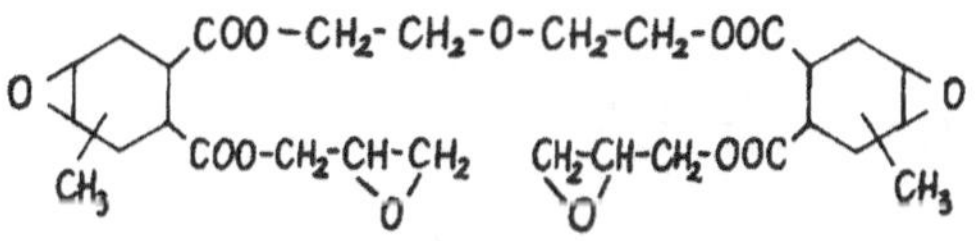

(LII)

Scheme 67

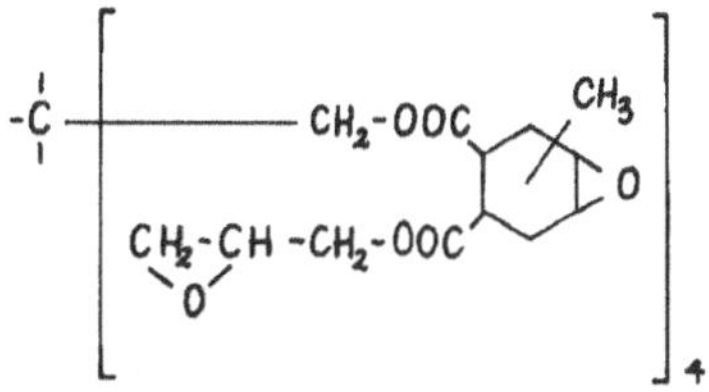

(LIII)

Scheme 68

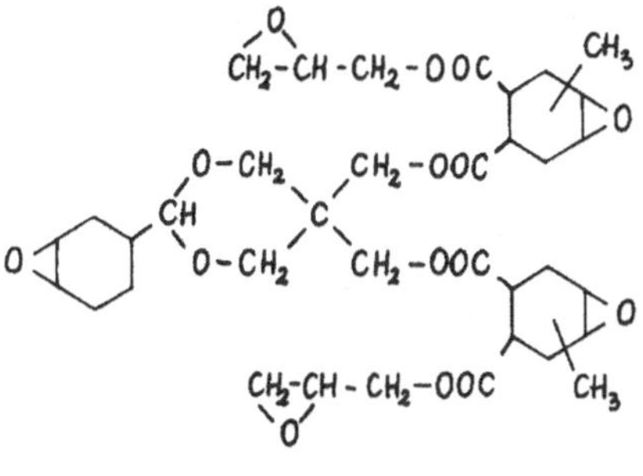

(LIV)

Scheme 69

Table 13. Properties of cured multifunctional ACECs with glycidyl ester groups and epoxycyclohexane rings

Epoxy compound		Epoxy groups		Properties of the cured epoxy compounds						
No.	Scheme	Glycidyl ester	Epoxy-cyclohexane	Curing agent	TM (°C)	Mechanical strength (MPa)			Elongation at break (%)	
						Tensile	Flexural	Compression		
L	65	1	2	AA MOCA	170 114	49 84	83 98	162 215	1.6 3.0	
LI	66	2	2	AA MOCA	149 117	57 92	76 105	155 182	2.2 3.0	
LII	67	2	2	AA m-PDA	120 97	73 85	92 123	131 143	2.7 3.2	
LIII	68	4	4	AA	162	44	67	155	2.1	
LIV	69	1	3	AA DDM	154 137	62 77	89 118	142 175	2.7 2.3	
LV	70	2	3	AA DDM	167 122	69 72	88 113	157 168	2.3 2.7	
LVI	71	1	2	AA DDM	130 122	64 74	85 98	159 143	3.5 3.8	
LVII	72	2	3	AA	140	67	92	165	3.0	
LVIII	73	1	4	AA DDM	169 137	59 63	79 114	166 161	3.2 2.6	

AA: 4-methyl-4-cyclohexene-1,2-dicarboxylic acid anhydride
m-PDA: m-phenylenediamine
DDM: 4,4'-diaminodiphenylmethane
MOCA: 3,3'-dichloro-4,4'-diaminodiphenylmethane

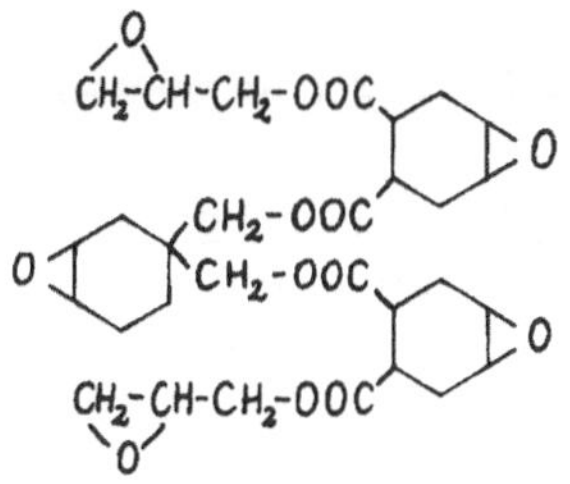

(LV)

Scheme 70

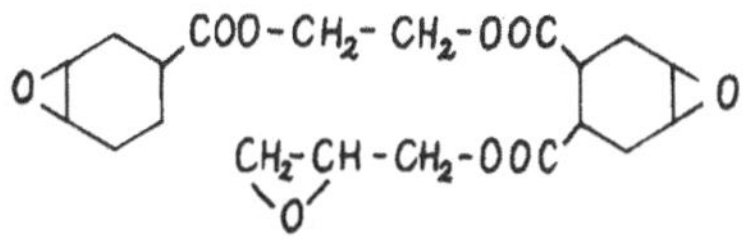

(LVI)

Scheme 71

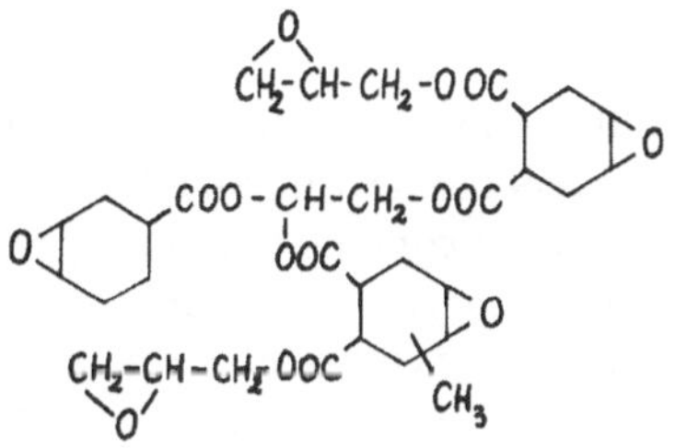

(LVII)

Scheme 72

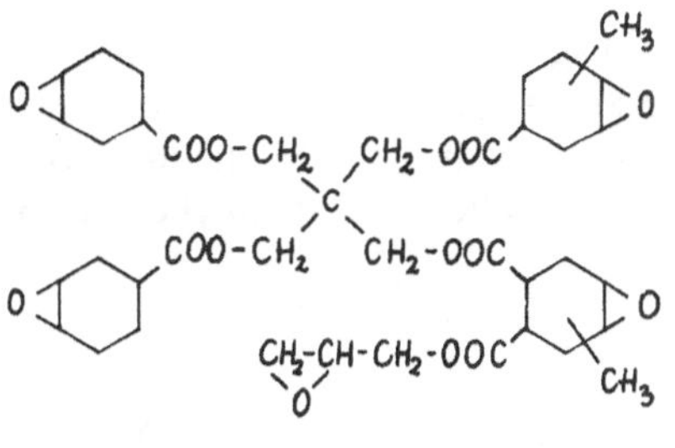

(LVIII)

Scheme 73

structural defects. Eventually, the mechanical strength properties and – in many cases – heat resistance are adversely affected. The unfavourable influence of the extended structure of ACECs is stronger in the case of amine curing in comparison with crosslinking with acid anhydrides (Table 13).

4.3
Curing of Aliphatic-Cycloaliphatic Epoxy Compounds with Phenol-Formaldehyde Resins

Crosslinked compositions consisting of epoxy compounds or resins and phenol-formaldehyde oligomers (PFOs) are distinguished by a combination of high mechanical strength and heat resistance as well as good dielectric properties [56]. The reaction mechanism is rather complicated because the system involves different reactive functional groups. Many investigations have been devoted to a study of the kinetics of the reaction of glycidyl epoxy groups with phenols [57–60]. However, there has been little information published about the reaction of cycloaliphatic epoxy groups with phenols. The investigations were limited to the non-catalyzed process [61]. An article about the reaction of ACECs with phenol was published in 1989 [62].

Investigation of curing ACECs with PFOs started from the study of the reactions of model monoepoxide compounds. Two epoxycyclohexane derivatives (**LIX** and **LX**, Schemes 74 and 75), one glycidyl ether (**XVIII** Scheme 18) and one glycidyl ester (**XVII**, Scheme 17) were used. Properties of the model monoepoxides are given in Table 14.

Phenol was applied as the model compound representing PFOs.

The reaction of the monoepoxides with phenol was followed at 120 °C in melt by determining the epoxy groups content. Equimolar ratios of the reagents were used. Oxalic acid, N,N-dimethylbenzylamine and $ZnCl_2$ (1% of the monoepoxide) were applied as catalysts. It was found that the reaction rates of the glycidyl and the epoxycyclohexane derivatives differ substantially (Fig. 8).

The increased reactivity of the cycloaliphatic epoxy group in the compound **LIX** in the non-catalyzed reaction with phenol results from the enhanced basicity of the epoxy oxygen in the epoxycyclohexane group. The incorporation of a carbonyl group (ester linkage in **LX**) decreases the electron density of the epoxycyclohexane oxygen and consequently decreases the reaction rate. The reaction rate of the cycloaliphatic epoxy group with phenol becomes close to that of the epoxy group in glycidyl 3-cyclohexene-1-methyl ether **XVIII**.

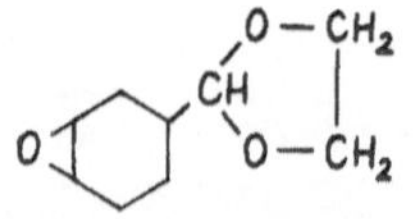

(LIX)

Scheme 74

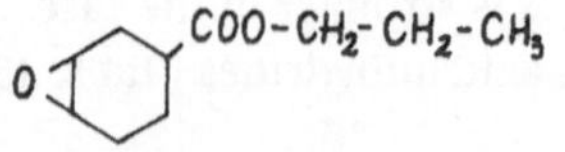

(LX)

Scheme 75

Table 14. Model monoepoxy compounds

Compound	Formula	Mol. weight	Elementary analysis (%)				Epoxy oxygen content(%)		Boiling temperature °C//Pa	Refractive index n_D^{25}
			Calc.		Found		Calc.	Found		
			C	H	C	H				
1-(2,5-Dioxolan-1-yl)-3,4-epoxy-cyclohexane	**LIX** $C_9H_{14}O_3$	170	63.55	8.29	63.49	8.24	9.40	9.28	105//20	1.4790
n-Propyl 3,4-epoxy-cyclohexane-1-carboxylate	**LX** $C_{10}H_{16}O_3$	184	65.19	8.75	65.12	8.70	8.70	8.63	126//20	1.4572
3-Cyclohexenyl-1-methyl glycidyl ether	**XVIII** $C_{10}H_{16}O_2$	168	71.39	9.58	71.20	9.53	9.52	9.29	129//20	1.4759
Glycidyl 3-cyclohexene-3-carboxylate	**XVII** $C_{10}H_{14}O_3$	182	65.91	7.74	65.50	7.50	8.79	8.56	132//30	1.4875

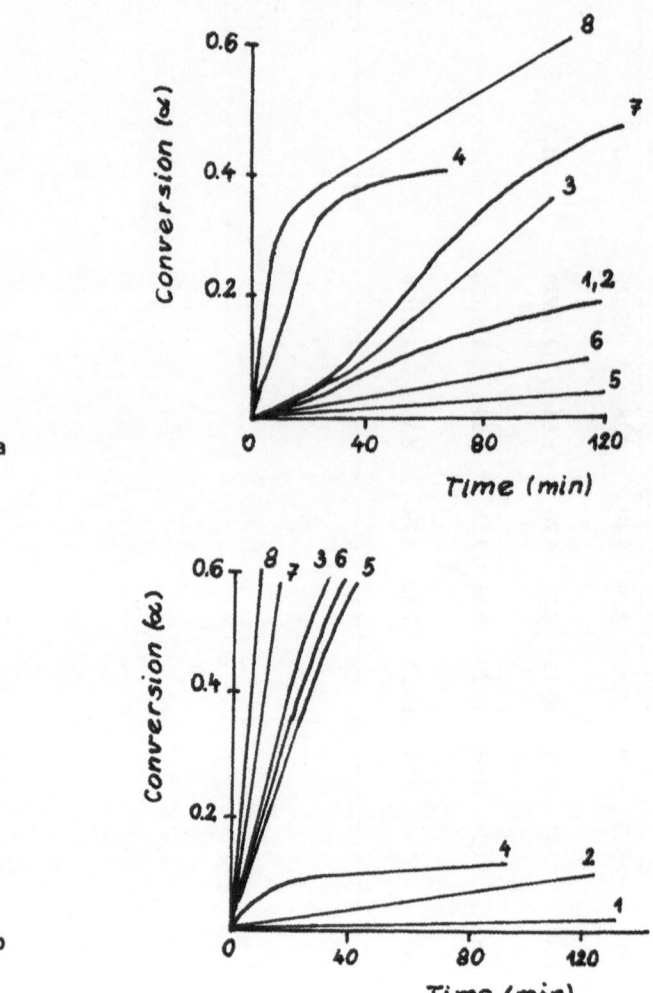

Fig. 8 a–b. Conversion (α) of model epoxy compounds with phenol at 120 °C in presence of 1% catalyst. **a** Cycloaliphatic epoxy compounds **LIX** and **LX**; (b) Glycidyl derivatives **XVIII** and **XVII** (see Table 14). **b** 1 – **LIX**, no catalyst; 2 – **LIX**, oxalic acid; 3 – **LIX**, *N,N*-dimethyl-benzylamine; 4 – **LIX**, ZnCl$_2$; 5 – **LX**, no catalyst; 6 – **LX**, oxalic acid; 7 **LX**, *N,N*-dimethyl-benzylamine; 8 – **LX**, ZnCl$_2$. (b) 1 – **XVIII**, no catalyst; 2 – **XVIII**, oxalic acid; 3 – **XVIII**, *N,N*-dimethylbenzylamine; 4 – **XVIII**, ZnCl$_2$; 5 – **XVII**, no catalyst; 6 – **XVII**, oxalic acid; 7 – **XVII**, *N,N*-dimethylbenzylamine; 8 – **XVII**, ZnCl$_2$

The high reactivity of glycidyl 3-cyclohexene-1-carboxylate **XVII** cannot be explained in terms of the basicity of the epoxy oxygen. The basicity in this case is much lower than that of cycloaliphatic epoxy oxygen due to the inductive acceptor effect of the carbonyl group. The explanation should be sought rather in a difference of the mechanism of the non-catalyzed reaction of phenol with epoxy groups in glycidyl esters in relation to epoxycyclohexane derivatives and glycidyl ethers as well [63].

Addition of a basic catalysts (*N,N*-dimethylbenzylamine) results in a considerable increase in the reaction rate of phenol with both cycloaliphatic and glycidyl epoxy groups. The rise of reactivity of 3-cyclohexene-1-methyl glycidyl ether **XVIII** is especially strongly pronounced. The catalytic role of the tertiary amine consists in an increase in the nucleophilic activity of phenol as the proton donor as a result of proton withdrawal.

Similar relationships were found when $ZnCl_2$ was used as a catalyst. The catalytic activity is generally higher in comparison with that of the tertiary amine. This statement is true for the cycloaliphatic epoxy groups and for the glycidyl ester whereas the catalytic effect of $ZnCl_2$ on the reaction of phenol with the glycidyl ether is very weak. A comparison of the course of the epoxide conversions (the curves in Fig. 8a,b) shows that the catalytic activity of oxalic acid is negligible.

The investigations on model compounds confirmed that it was possible to carry out curing of ACECs or compositions of cycloaliphatic epoxy compounds with glycidyl esters or ethers, so that the different epoxy groups were consecutively exhausted. The consecutive entering of various epoxy groups into the reaction with phenolic hydroxyls can take place independently of the catalyst being used. Such a course of reaction can presumably contribute to a more regular structure of the ACEC/PFO polymer networks.

The curing with PFOs can be replaced by curing with phenol-aniline-formaldehyde oligomers (PAFOs) which contain reactive amino groups. The replacement results in an increase in mechanical strength and heat resistance of composites (Table 15) thanks to an enhanced reactivity of the system, shortening of the curing time and higher conversion of the functional groups. Nevertheless, an increase in the mechanical strength and heat resistance of the composites can also be achieved by an addition of 1% $ZnCl_2$ to the epoxy compounds cured with PFO. Further improvement of some properties of the PAFO cured systems is also achieved by adding of $ZnCl_2$. The catalyst does not improve the mechanical strength of the composite at room temperature. However, the addition of $ZnCl_2$ results in a considerable increase in the flexural strength at 200 °C thanks to the formation of a polymer matrix with high crosslinking density.

The triepoxide ACECs exhibit flexural strength values at 200 °C which are distinctly higher than those of the diepoxy compounds. This difference can be attributed to the higher functionality of the triepoxides and, presumably, to the improved regularity of the polymer network. Nevertheless, there is no direct evidence for the network regularity.

Table 15. Properties of glass fiber reinforced epoxy/PFO and epoxy/PAFO plastics

Epoxy compound (UP-)	Curing agent	Flexural strength (MPa)			
		at 25 °C		at 200 °C	
		no catalyst	with 1% ZnCl$_2$	no catalyst	with 1% ZnCl$_2$
640	PAFO-1	835	860	94	140
	PFO	830	840	60	125
640T	PAFO-1	837	700	128	280
	PFO	800	800	90	180
650	PAFO-1	910	860	94	165
	PFO	780	820	56	125
650T	PAFO-1	923	910	347	450
	PFO	800	820	250	350

Characteristics of the epoxy compounds (see Table 10):
UP-640:   diglycidyl ester, **XIX** (Scheme 19)
UP-640T: ACEC triepoxide with glycidyl ester linkages, **XLVIII** (Scheme 63)
UP-650:   diglycidyl ether, **LIX** (Scheme 74)
UP-650T: ACEC triepoxide with glycidyl ether linkages, **XXXV** (Scheme 39)

Noteworthy is the extremely high flexural strength at 200 °C of the PAFO cured triepoxide UP-650 T (one cycloaliphatic epoxy group and two glycidyl ether groups) with 1% ZnCl$_2$ catalyst.

Investigation of curing of the triepoxide ACECs and their diepoxide analogues with PFO and PAFO by means of the electrokinetic method (Fig. 9) confirmed the principal conclusions drawn from the data presented in Fig. 8. The compounds containing cycloaliphatic epoxy groups exhibit a decreased curing rate and a decreased conversion at 120 °C. However, the curing process can be completed by post-curing for 4 h at 150 °C or for 2 h at 180 °C, if the PAFO is combined with the glycidyl ether derivatives UP-650 and UP-650 T. The curing of glycidyl esters UP-640 and UP-640 T with PAFO is completed only after 2 h at 200 °C.

It was shown that the conversion of epoxy groups on curing with PFO is incomplete even after post-curing for 4 h at 200 °C: some unreacted cycloaliphatic epoxy groups were detected at 820 cm^{-1} by means of IR spectroscopy.

The addition of ZnCl$_2$ shifts the range of the intense curing process towards lower temperatures.

Properties of the commercial composites are given in Table 16. The polymer matrix is based on epoxy compounds with glycidyl ether groups (one of the compounds is a triepoxide ACEC). The PAFO used in this case differs from that reported in Table 15 in the amine and hydroxymethyl groups content. Properties of the composites prepared from corresponding diglycidyl esters (one of them containing, moreover, a cycloaliphatic epoxy group) are presented in Table 17. The composites prepared from the triepoxides (both glycidyl ether and ester

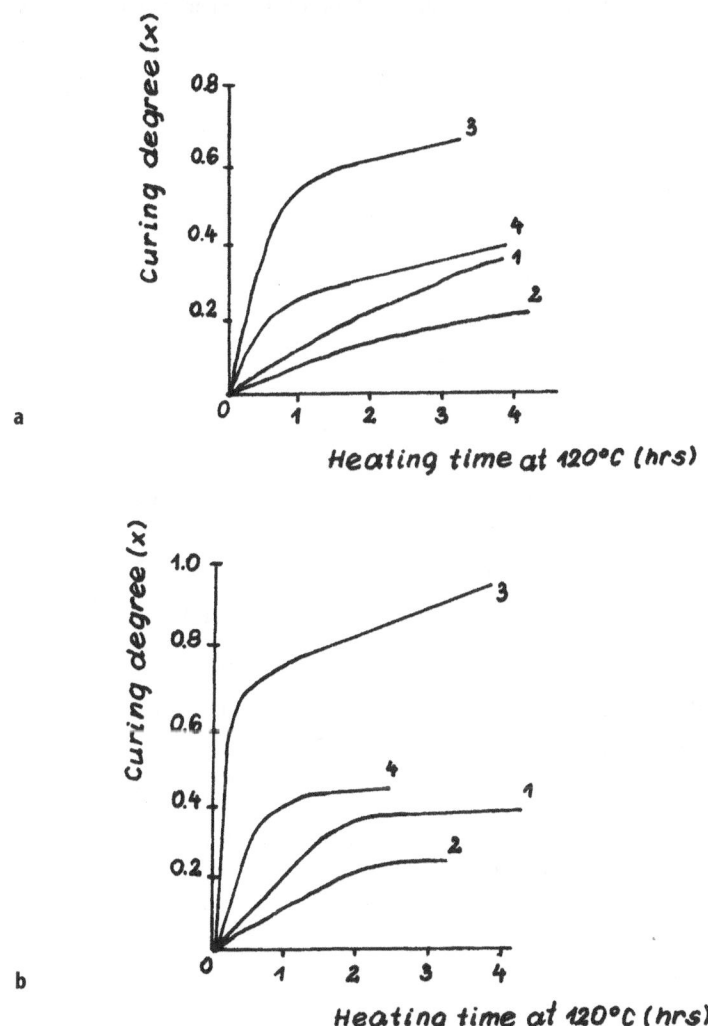

Fig. 9 a-b. Time dependence of the curing degree (X) of epoxy compounds with a PFO and a PAFO at 120 °C: **a** curing with PFO; **b** curing with PAFO. 1 – UP-640; 2 – UP-640 T; 3 – UP-650; 4 – UP-650 T (see Table 15)

type) and the PAFO curing agents exhibit high flexural strength at 200 °C. The results given in Table 17 show that higher strength values are obtained with the carbon fiber reinforcement. It can be assumed that this advantegeous property results from high adhesion between the carbon fibers and the polymer matrix. The high adhesion is presumedly caused by the reaction between carboxyl and hydroxyl groups being present on the surface of the carbon fibers and the epoxy groups in ACECs. The cycloaliphatic epoxy groups are particularly reactive.

Table 16. Mechanical properties of composites made of the PAFO cured epoxy compounds (glycidyl ether type) [64]

Epoxy compound	UP–650 (LIX)		UP–650T (XXXV)	
Curing agent	PAFO–2		PAFO–2	
Epoxy groups/phenolic OH ratio	1:2.5		1:2	
Reinforcement	"1"	"2"	"1"	"2"
Flexural strength (MPa)				
at 25 °C	845	790	785	810
at 200 °C	160	450	560	530

"1" – glass cloth,
"2" – carbon fiber tape

Table 17. Mechanical properties of composites made of the PAFO cured epoxy compounds (glycidyl ester type)

Epoxy compound	UP–640 (LVIII)				UP–640T (XLVIII)			
Curing agent	PAFO–1		PAFO–2		PAFO–1		PAFO–2	
Epoxy groups/phenolic OH ratio	1:1.5		1:1		1:2		1:2.5	
Reinforcement	"1"	"2"	"1"	"2"	"1"	"2"	"1"	"2"
Flexural strength (M Pa)								
at 25 °C	837	965	858	969	895	980	945	990
at 200 °C	90	300	300	280	270	435	420	444

PAFO–1 was used in Table 15 whereas PAFO–2 was used in Table 16
"1" – glass fiber,
"2" – carbon fiber

4.4
Composites Based on Aliphatic-Cycloaliphatic Epoxy Compounds

Incorporation of aromatic rings into the molecules of ACECs should result in an increase in thermal stability. Several ACECs were synthesized comprising glycidyl ether substituted phenols and bisphenols with built in epoxycyclohexane rings (**XLIII, LXI** and **LXII**; Schemes 50, 76, and 77) [65–68]. The ACECs **LXIV** were synthesized by the esterification of 3-cyclohexenyl-1-methanol with salicylic, β-resorcylic or gallic acid followed by the glycidylation of phenolic groups with ECH and epoxidation of the cyclohexenyl ring with PAA [68, 69]. After curing, they exhibit particularly high mechanical strength, a satisfactory flexibility and elevated heat resistance and thermal stability.

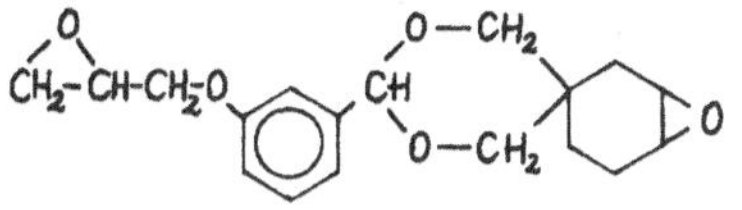

(LXI)

Scheme 76

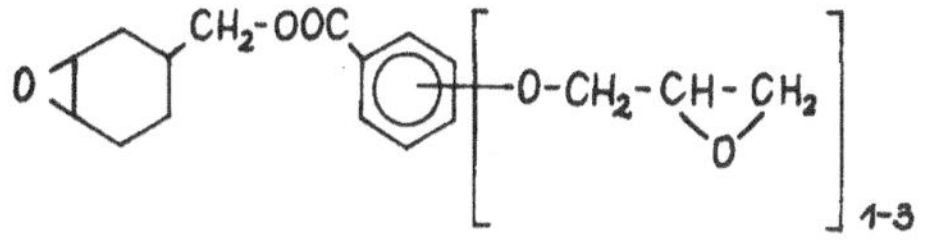

(LXII)

Scheme 77

Mechanical and thermal properties of the ACEC **XLIII** cured with an acid anhydride (cast profiles) and with PAFOs (a reinforced plastic) are compared with the properties of an epoxynovolak resin cured with the same curing agents (Tables 18 and 19). In addition to the high values of flexural strength at 200 °C, good thermal stability (weight loss of only 1.8% after 500 h at 200 °C) was found.

Table 18. Properties of ACEC XLIII and an epoxynovolak resin cured with an acid anhydride

Curing agent	4-Methyl-4-cyclohexene-1-carboxylic acid anhydride	
Epoxy compound	ACEC **XLIII**	Epoxynovolak resin
TM	160	190
Flexural strength (MPa)	100	100
Impact strength(kJ m^2)	17	8
Temperature of the start of thermal decomposition	260	245

Table 19. Properties of composites made of the ACEC **XLIII** and an epoxynovolak resin (UP–643) cured with PAFOs

Epoxy component	ACEC **XLIII**		Epoxynovolak resin	
Curing agent	PAFO–1	PAFO–2	PAFO–1	PAFO–2
Epoxy groups/phenolic OH ratio	1:1	1:2	1:1	1:2.5
Flexural strength (MPa)				
at 25 °C	850	760	888	858
at 200 °C	350	450	120	300

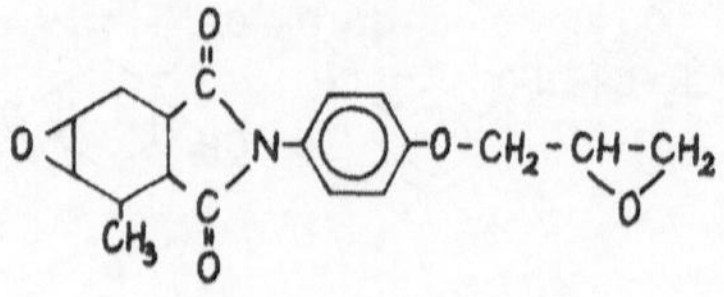

(LXIII)

Scheme 78

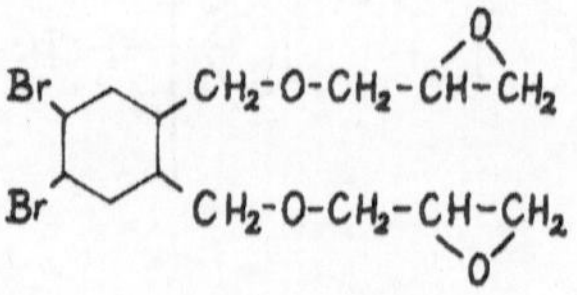

(LXIV)

Scheme 79

It is well known that polymers with condensed imide rings exhibit high thermal stability. Properties of the imide ring containing ACEC **LXII** (Scheme 78), cured with amines and with an acid anhydride, are given in Table 20. Acid anhydride cured epoxyimides reach the heat resistance of 190 °C, temperature of the start of decomposition above 265 °C and flexural strength above 70 MPa.

An epoxyimide cured with a PFO was used as binder for glass fiber reinforced composites. The composites are characterized by the TM value above 260 °C and flexural strength up to 600 MPa. They exhibit outstanding dielectric properties which are stable during long-term heating. Such polymers and composites are particularly suitable for long-term uses at elevated temperatures (above 200 °C). The epoxyimides are superior to the usual CECs with epoxycyclohexane rings: their maximum long-term working temperature does not exceed 200 °C [70–72]. These findings were generally confirmed in articles concerning ACECs with a different chemical structure [4, 73].

Properties of an amine and acid anhydride cured brominated epoxy compound **LXIV** (Scheme 79) which contains a cycloaliphatic ring are presented in

Table 20. Properties of the imide ring containing ACEC **LXIII** cured with various hardeners

Curing agent	TM (°C)	Mechanical strength (MPa)		Elongation at break (%)
		Tensile	Compression	
m-Phenylenediamine	140	116	160	5.0
4,4'-Diamino-3,3'-dichlorodiphenylmethane	135	109	147	5.2
4,4'-Diaminodiphenylmethane	150	109	130	7.4
4-Methyl-4-cyclohexene-1,2-dicarboxylic acid anhydride	161	100	130	6.1

Table 21. Properties of cured 3,4–dibromo–1,1–di(2,3–epoxypropoxymethyl)cyclohexane (LXIV, Scheme 79)

Curing agent	TM (°C)	Mechanical strength (MPa)			Elongation at break (%)	Self-extinguishing time (s)
		Tensile	Flexural	Compression		
m–Phenylene diamine	165	60	120	140	3.0	1
4–Methyl–4–cyclohexene–1,2–dicarboxylic acid anhydride	200	50	–	170	2.0	1

Table 21. Highs level of the parameters characterizing the mechanical and thermal properties as well as good processing properties show the superiority of the compound **LXIV** to the tetrabromobisphenol A based epoxy resins.

5
Modification of Epoxy Compounds and Resins with Aliphatic-Cycloaliphatic Epoxy Compounds

5.1
Compositions Consisting of Cycloaliphatic and Aliphatic-Cycloaliphatic Epoxy Compounds

The structure of polymer networks which are formed as a result of crosslinking of epoxy resins and compounds is determined by the weight ratio and the structure of the epoxy component and the curing agent, by the sequence of entering of the individual functional groups into the reaction, by the conversion degree of the functional groups, etc. Properties of the crosslinked polymers depend, moreover, upon the strength of Van der Waals forces and hydrogen bonds, of the polymer chain entanglement and tacking, density of the covalent bonds, crosslinking density, etc. [41, 46].

It has been shown that the modification of CECs and epoxynovolaks with monoepoxides results in an improvement of properties of the crosslinked polymers. Under certain conditions, the monoepoxides eliminate the structural defects, thus equalizing density and increasing the mechanical strength and heat resistance [74].

The modifying performance of cycloaliphatic monoepoxides in the curing processes of cycloaliphatic diepoxy compounds was described in [75]. ACECs are very efficient as modifiers of epoxyamine [76] and epoxynovolak [77–79] resins. The search for reactive diluents for epoxy compositions was initially directed towards an improvement of the processing characteristics, in particular the viscosity suppression, whereas the increase in mechanical strength and heat resistance was considered to be the result of the improved processing character-

Table 22. Effect of the ACEC modifiers on some properties of CECs [82]

ACEC modifier (UP–)	Compound	Dicyclopentadiene dioxide (DCPDO) based compositions				Tetrahydroindene dioxide (THIDO) based compositions	
		DCPDO/ACEC weight ratio				THIDO/ACEC weight ratio	
		1:1		2:1		1:1	
		RR	T_g (°C)	RR	T_g (°C)	RR	T_g (°C)
–	–	7	210	4	215	6	195
640	XIX	15	200	10	220	16	101
640T	XLVIII	8	260	7	270	9	183
650	LIX	16	160	8	250	13	123
650T	XXXV	8	293	4	305	9	196

RR – relative reactivity
Curing agent: a 50:70 by weight mixture of maleic anhydride with 4-methyl-4-cyclohexene-1,2-dicarboxylic acid anhydride

istics. It should be stressed that the addition of ACECs as reactive diluents usually results in an increase in the glass transition temperature [79].

The following criteria should be taken into account in the selection of novel reactive modifiers [74]:

– low molecular weight, close to that of the epoxy compounds to be modified,
– low viscosity and high degree of homogeneity of the modified epoxy compositions under the conditions of processing,
– the presence of various epoxy groups, differing in reactivity.

Many ACECs fulfill these requirements better than the glycidyl ethers which are commonly applied as reactive diluents. The molecular weights of ACECs are usually a little higher than those of the monoglycidyl ethers (butyl or phenyl glycidyl ether, etc.), although the molecular weight fraction per one epoxy group of ACECs is considerably lower. Moreover, the triepoxide ACECs add one more branching to the polymer network. The increase in the crosslinking density which results from the replacement of the monoglycidyl ethers as reactive diluents by ACECs should eventually cause an improvement of the functional quality [80, 81]. The reality of such an approach is exemplified by the investigation of the properties of the ACECs modified CECs. For that study, CECs with the most rigid structure – dicyclopentadiene dioxide (DCPDO) and tetrahydroindene dioxide (THIDO) – were applied (Tables 22–24].

Both epoxy groups in THIDO differ only a little from each other in reactivity. Consequently, the advantageous effect of the ACEC modifiers is not as strong as in the case of DCPDO which is characterized by a considerable difference between the reactivity of the epoxy group attached to the cyclopentane ring and that attached to the norbornane (endomethylene-cyclohexane) ring.

Table 23. Properties of acid anhydride cured and ACEC modified dicyclopentadiene dioxide (DCPDO) – DCPDO/ACEC (or the epoxy resin), weight ratio 2:1

ACEC modifier (UP–)	TM (°C)	Mechanical properties				Water absorption 24 h (%)
		Flexural strength (MPa)		Impact strength (kJ m^{-2})	Brinell hardness (MPa)	
		at 25 °C	at 200 °C			
640	195	77	39	8	17.0	0.41
640T	210	95	40	12	15.6	0.41
650	189	66	41	14	17.0	0.34
650T	226	92	71	10	16.2	0.31
BPA/ECH epoxy resin	160	110	50	15	–	–

Curing agent: see Table 22

Table 24. Properties of acid anhydride cured and ACEC modified tetrahydroindene dioxide (THIDO) –THIDO/ACEC (or the epoxy resin) weight ratio 2:1

ACEC modifier (UP–)	Mechanical strength (MPa)		Elongation at break (%)	Mechanical strength decrease at 150 °C (%)
	Tensile	Flexural		
640	57	84	2.2	57
640T	56	90	2.2	10
650	66	96	3.4	46
650T	62	120	4.6	11
BPA/ECH epoxy resin	61	70	1.7	69

Curing agent: see Table 22.

The high modifying efficiency of ACECs in the DCPDO and THIDO based compositions afforded the authors a handle to an examination of the effect of ACEC modifiers on the properties of a larger group of CECs cured with acid anhydrides. General conclusions from that investigation are that the application of ACECs in the CEC based compositions makes it possible to improve the processing characteristics and, in particular, to combine an increase in mechanical strength with an increase in heat resistance. As stated before, the effect achieved may be attributed to an enhanced structural regularity and crosslinking density of the polymer networks, thanks to the different reactivity of the cycloaliphatic and the glycidyl epoxy groups in ACECs [41, 46, 83].

Table 25. Mechanical properties of ester and cyclic acetal type CECs modified with ACECs

CEC diepoxide (UP–)	ACEC modifier (UP–)	Mechanical strength(MPa)		Elongation at break (%)
		Tensile	Flexural	
612	–	45	70	2.0
	650	50	75	2.5
XXVI	650T	63	83	3.0
	640	54	78	2.6
(Scheme 28)	640T	62	87	2.8
650	–	50	80	2.0
	650	57	100	3.0
XXIV	650T	76	117	6.0
	640	59	98	3.1
(Scheme 26)	640T	74	110	4.8

Curing agent: see Table 22.

Table 26. Elongation at break at elevated temperatures of CECs modified with triepoxide ACECs

CEC diepoxide (UP–)	ACEC modifier (UP–)	Temperature of the tensile test (°C)	Elongation at break (%)			
			Modifier content (%)			
			10	20	30	40
612	650T	190	17	20	17	17
(**XXVI**)	640T	190	12	14	15	18
632	650T	130	13	18	16	17
(**XXIV**)	640T	130	15	18	16	17

Curing agent: see Table 22.

The properties of the cured CEC/ACEC compositions are affected not only by the chemical structure of the ACEC modifiers but also by the character of the linkages between the epoxycyclohexane rings in CECs. An increase in the mechanical strength and elongation at break by the addition of ACECs is much more distinct in the case of the CEC with ester linkage (UP-632, **XXIV**, Scheme 26) in comparison with the CEC with the cyclic acetal linkage (UP-612, **XXVI**, Scheme 28) (Table 25). This difference is due to the greater mobility of cyclohexane rings linked with ester bonds.

The rise of the testing temperature results in a decrease in the influence of intramolecular forces on the mechanical strength and deformability of polymers. Consequently, the elongation at break at elevated temperatures becomes nearly independent from the ACECs content in a broad concentration range (Table 26).

5.2
Composites Prepared from Cycloaliphatic and Aliphatic-Cycloaliphatic Epoxy Compounds Crosslinked with Phenol-Formaldehyde Oligomers

Phenolic hydroxyls in phenol-formaldehyde resols (resol type PFOs) react with epoxy groups in ACECs. The reaction occurs with epoxy groups in both cycloaliphatic and glycidyl units. Additional crosslinking results from the condensation of two CH_2OH groups in the PFO resol. The phenolic OH/epoxy groups addition and the CH_2OH condensation differ from each other as far as the energetic levels are concerned and are not competitive. Consequently, the crosslinking reactions can take place step by step, resulting in an increase in mechanical strength. In this respect, the results of [84] related to the modifying efficiency of ACECs in the PFO resols cured CECs should be mentioned.

Flexural strength values determined at 25 and 200 °C of glass cloth reinforced CEC/PFO systems are given in Table 27. The strength at 25 °C is close to that of similar composites with PFO cured BPA/ECH epoxy resins. However, the flexural strength at 200 °C of the CECs-based materials is several times higher. This

Table 27. Glass fiber reinforced CECs cured with a PFO

Cycloaliphatic diepoxy compound "CEC"	CEC/PFO weight ratio	Flexural strength (MPa)			
		Glass cloth "1"		Glass cloth "2"	
		at 25 °C	at 200 °C	at 25 °C	at 200 °C
UP–612	1:1	780	230	560	220
	1:1.5	800	270	590	250
UP–632	1:1	785	300	580	275
	1:1.5	785	290	610	250
DCPDO	1:1	665	–	510	320
	1:1.5	690	340	540	320
THIDO	1:1	670	–	530	240
	1:1.5	720	360	600	300
BPA/ECH epoxy resin	1:1.5	600	<200	–	–

Solutions of the binders in acetone were used to impregnate the glass cloth

difference can be attributed to the higher glass transition temperature of the CEC/PFO polymers.

The strength parameters of the glass cloth "2" reinforced plastics are inferior because they have a paraffin based finish which decreases the adhesion between glass fiber and the polymer matrix. Shelf life of the CEC/PFO binders and pre-pregs is a little shorter than that of the corresponding BPA/ECH epoxy resin-based materials [84, 85].

The optimum CEC/PFO ratio is in the range of 60:40 – 70:30.

The composites based on the PFO resols cured ester type (UP-632, **XXIV**) and cyclic acetal type (UP-612, **XXVI**) CECs exhibit higher mechanical strength and heat resistance in comparison with similar materials cured with dicarboxylic acid anhydrides [86].

The examination of CEC/PFO systems and the modification of the CEC/PFO systems with ACECs and with six-membered ring containing diglycidyl deriva-tives was aimed at an increase in mechanical strength with no deterioration of the high heat resistance. The cyclohexene ring containing diglycidyl ester (UP-640, **XIX**), glycidyl ethers (UP-650, **LIX** and UP-643, **XLII**) as well as the corre-sponding triepoxide ACECs prepared by the epoxidation of the cyclohexene units in the diglycidyl derivatives (**XLIII** from **XLII**, Scheme 50) with PAA were crosslinked with a PFO (Table 28).

The DCPDO-based crosslinked polymers and composites exhibit similar me-chanical strength and heat resistance in comparison with those prepared from UP-612, UP-632 and THIDO (Table 27). The increase in functionality of the di-glycidyl ether or ester modifiers by the epoxidation of the cyclohexene ring re-sults in a simultaneous increase in both mechanical strength and heat resistance

Table 28. Glass fiber reinforced THIDO modified with ACECs and cured with a PFO

ACEC modifier (UP–)	THIDO/ACEC weight ratio	Flexural strength (MPa)	
		at 25 °C	at 200 °C
650T	1:1	790	390
	2:1	835	497
650	1:1	880	310
	2:1	880	340
643T	1:1	850	465
	2:1	810	470
643	1:1	870	480
	2:1	820	465
640T	1:1	820	344
	2:1	850	440
640	1:1	820	300
	2:1	840	320

of the DCPDO/PFO polymers and composites (Tables 28 and 29). Particularly high flexural strength at 200 °C was achieved when the triepoxide UP-643 T (**XLIII**) was applied as the ACEC modifier (Table 29).

The systems based on THIDO exhibit higher reactivity than those based on DCPDO. This statement is based on the shorter gel time and higher growth rate of the gel fraction content on storage of the prepregs. As far as the systems with the CECs, modifiers (diluents) and PFO are concerned, an effect of the modifiers on the shelf life is observed. The triepoxide ACEC type diluents increase the shelf

Table 29. Glass fiber reinforced DCPDO modified with ACECs and cured with a PFO

ACEC modifier (UP–)	DCPDO/ACEC weight ratio	Flexural strength (MPa)	
		at 25 °C	at 200 °C
650T	1:0.5	470	320
	1:1	490	340
	1:1.5	920	470
643T	1:0.5	840	500
	1:1	740	440
	1:1.5	780	510
640T	1:0.5	710	340
	1:1	400	180
	1:1.5	320	95

Reinforcement: glass cloth

life, whereas the addition of diglycidyl ethers or esters, which contain no epoxy-cyclohexane rings, results in a decrease in shelf life.

Prolongation of the storage time of prepregs is very desirable. Therefore, an attempt was made to explain the differences by means of IR spectroscopy. It was found that at temperatures up to 100 °C the polycondensation of the PFO resol occurred with the participation of CH_2OH groups. The polycondensation results in a decrease in the functional groups content in the system and adversely affects the properties of the crosslinked polymer. The gel fraction of the polymers which were cured initially at 60–100 °C did not exceed 80–85% even after prolonged postcuring for 10 h at 200 °C. If the initial curing was carried out at 120–140 °C, the reaction of glycidyl groups and phenolic hydroxyls prevailed. The conversion of the cycloaliphatic epoxy groups is incomplete even after postcuring at 180–220 °C.

The prolonged shelf life of the binders and prepregs based on CECs, ACECs and PFOs results consequently from the lower reactivity of epoxy groups in epoxycyclohexane rings. Accordingly, the stepwise curing process was applied. The manufacturing process of the CEC/ACEC/PFO composites involved the increase in the curing temperature from 140 up to 220 °C under a pressure of 0.05–0.5 MPa.

The data presented show that a proper selection of modifiers and curing parameters makes it possible to develop heat resistant composites with high temperature resistance. The composites are superior to similar materials based on BPA/ECH epoxy and epoxynovolak resins.

Polymeric materials based on CECs are very rigid structures with high crosslinking density. Therefore, the heat deflection temperatures (TM) are higher than the temperatures of the start of thermal decomposition observed in the dynamic heating process [71].

The data characterizing the changes of the properties of the crosslinked polymers during aging in air (Table 30) show that the increase in heat deflection temperature (TM) and hardness is accompanied by a decrease in flexural strength. It is noteworthy that the modification with ACECs results in an increase in hardness, whereas the flexural strength does not change.

Table 30. Thermal aging resistance of DCPDO modified with ACECs and cured with a PFO, the properties after aging at 200 °C

ACEC modifier (UP–)	TM (°C)		Flexural strength (MPa)		Brinell hardness (MPa)	
	150 h	250 h	150 h	250 h	150 h	250 h
–	235	242	80	70	169	178
640T	225	230	58	50	197	221
643T	248	253	83	78	236	239

The parameters of the thermal decomposition process were determined by means of the thermogravimetric method. The decomposition starts at 248–252 °C. The activation energy of the thermal decomposition process is in the range of 98.4–101.1 kJ/mol. The weight loss up to 500 °C amounts to above 70% and the weight loss rate is 0.018–0.022%/s. The character of the thermogravimetric curves corresponds to a one-stage process. It is, however, rather improbable that polymers having such complicated structure decompose in a one-stage process. It can be assumed that the limiting stages of the multistage process have close kinetic parameters and therefore the whole course of decomposition is characterized by an average energy of activation. The calculated order of reaction is close to one.

6
General Remarks

The use of ACECs as modifiers for CECs makes it possible to preserve the desirable processing characteristics of epoxy resins and to improve the mechanical strength, elongation under load and heat resistance. The thermal stability of the crosslinked materials does not depend on the type of modifier. This statement concerns both the thermooxidative aging in the isothermal and the dynamic conditions.

The general conclusions were confirmed by the practical results related to the co-curing of CEC/ACEC compositions with amines in the presence of BF3-amine complexes [87]. The idea presented here was also applied to the synthesis of novel epoxy compounds which contained epoxy groups differing in structure and reactivity [88–90] to be used in adhesive formulations.

Some related problems are also discussed in a review [91].

A detailed presentation of all published and unpublished data concerning ACECs and CEC-ACEC compositions is almost impossible within a framework of a review having a limited size. Nevertheless the authors hope to have described the ways of improving the properties of a definite class of polymeric materials by means of an appropriate synthesis of monomers characterized by a differentiated structure of various epoxy groups in the molecule.

7
References

1. Brojer Z., Hertz Z, Penczek P (1972, 1982) Żywice epoksydowe (Epoxy resins), 2nd and 3rd edn. WNT, Warsaw
2. Brojer Z, Penczek P, Penczek S (1962) Przemysl Chem. (Warsaw) 41:437, 684
3. Moshinski L (1995) Epoksidnye smoly i otverditeli (Epoxy resins and curing agents). Arkadia Press, Tel Aviv
4. Glukhen'ka LF et al. (1973) Plast Massy (12):55
5. Lynch BM, Pausacker KH (1955) J Chem Soc 1525
6. Schwartz NN, Blumbergs JH (1964) J Org Chem 29:1976
7. Sapunov VN, Lebedev NN (1966) Zhurn Org Khim 2:225

8. Sapunov VN, Lebedev NN (1965) Izv Vysshikh Uchebn Zavedeni, Khim Khim Tekhnol 8:771
9. Metelina DI (1972) Usp Khim 41:1737
10. Curci R (1970) J Org Chem 35:740
11. Bartlett PD (1950) Record Chem Progress 11:47
12. USSR Pat 1,513,866
13. Inglis DB (1971) Chem Ind (London) 44:1268
14. Rouchaut IZ (1972) Ind Chim Belge 37:741
15. Metelina DI, Shibaeva LV (1972) Neftekhimiya 12:2160
16. Batog AE et al. (1981) Zhurn Org Khim 17:1101
17. Kunčický I., Černý O (1973) Chem Prumysl 23:454
18. Batog AE, Batrak TA, Savenko TV (1980) unpublished results
19. Batzer H (1966) Makromol Chem 91:195
20. Kozlova LV, Batog AE (1975) unpublished results
21. Swiss Pat 480,323
22. Ger. Pat 1,211,177
23. USSR Pat 431,162
24. Malinovski MS (1961) Okisi olefinov i ikh proizvodnye (Olefin oxides and their derivatives). Goskhimizdat, Moscow
25. USSR Pat 516,722
26. Stepko OP et al. (1976) Plast Massy (4):69
27. USSR Pat 525,862
28. USSR Pat 466,257
29. Batog AE, Kiryushina NP (1975) unpublished results
30. Batog AE et al. (1985) unpublished results
31. Siling MM et al. (1969) Vysokomol Soed A 11:1943
32. Artemov VN et al. (1980) Plast Massy (10):17
33. Batog AE et al. (1979) Plast Massy (10):6
34. USSR Pat 514,823
35. USSR Pat 649,718
36. Gefter EL, Kabachnik MI (1962) Usp Khim 31:285
37. USSR Pat 449,077
38. USSR Pat 639,882
39. USSR Pat 759,516
40. USSR Pat 759,516
41. Andrianov NA, Emel'yanov VN (1976) Usp Khim 45:1817
42. Durmanenko NA et al. (1973) Plast Massy (3):45
43. Pet'ko IP, Batog AE (1976) Plast Massy (2):39
44. Pet'ko IP et al. (1974) Plast Massy (5):60
45. Paramonov Yu M, Pet'ko IP, Batog AE (1979) unpublished results
46. Lipatova TE (1974) Catalytic polymerization of oligomers and the formation of polymer networks. Naukova Dumka, Kiev (in Russian)
47. Firsov VA, Liskova EM, Pet'ko IP (1983) Plast Massy (9):53
48. Batog AE, Kozlova LV (1982) unpublished results
49. Pakter MK, Batog AE (1982) unpublished results
50. Pakter MK et al. (1981) Dokl Akad Nauk Ukr SSR, Ser B (8):48
51. Batog AE et al. (1980) Zhurn Org Khim 16:1126
52. Pet'ko IP et al. (1984) Plast Massy (5):47
53. Nechitaylo LG et al. (1987) Kinet Katal 28:1322
54. Pet'ko IP et al. (1981) Plast Massy (1):30
55. Pet'ko IP, Beida VI, Batog AE (1980) Plast Massy (5):18
56. Lapitskii VA, Kricuk AA (1986) Physicomechanical properties of epoxy polymers and glass fiber reinforced plastics. Naukova Dumka, Kiev (in Russian)
57. Lebedev NN, Shvets VF (1965) Kinet Katal 6:782

58. Sorokin MF, Shode LG (1966) Zhurn Org Khim 2:1463
59. Sheblanova MA et al., (1976) Dokl Akad Nauk SSSR 226:390
60. Golubok Yu A et al. (1979) Zhurn Org Khim 15:2106
61. Batog AE et al. (1984) unpublished results
62. Pet'ko IP et al. (1989) Zhurn Prikl Khim (Leningrad) 62:642
63. Klebanov MS, Kiryazev F Yu (1984) Zhurn Org Khim 20:2407
64. Pet'ko IP, Beida VI, Batog AE (1980) Plast Massy (5):18
65. Batog AE et al. (1979) unpublished results
66. Batog AE, Kryushina NP (1975) Epoksidnye Smoly i Materialy na Ikh Osnove, Trudy (NPO "Plastik", Moscow) 1:31
67. Tkachuk BM et al. (1985) Reaktsyonnosposobnye Oligomery i Kompozitsonnye Materialy na Ikh Osnove, Trudy (NIITEKhIM, Moscow): 3
68. USSR Pat 591,471
69. Artemov VN, Beida VI, Pet'ko IP (1980) Plast Massy (10):17
70. Pet'ko IP, Batog AE (1976) Plast Massy (10):6
71. Gurenko LV, Batog AE, Pet'ko IP (1979) Plast Massy (9):51
72. Pet'ko IP, Batog AE, Zaytsev Yu S (1987) Kompoz. Polim. Mater. 34:10
73. Georgitsa TA, Batog AE, Pet'ko IP (1988) Plast Massy (5):17
74. Pet'ko IP et al. (1986) Kompoz Polim Mater 29:47
75. Batog AE, Savenko TV, Pet'ko IP (1980) Plast Massy (8):44
76. Beida VI, Artemov VN, Pet'ko IP (1980) Plast Massy (7):58
77. Pet'ko IP et al. (1981) Plast Massy (2):58
78. Pet'ko IP, Beida VI, Batog AE (1980) Plast Massy (5):18
79. Pet'ko IP, Batog AE, Savenko TV (1982) Plast Massy (2):53
80. Pet'ko IP et al. (1981) Plast Massy (11):29
81. Pet'ko IP, Batog AE, Stepko OP (1985) Kompozitsionnye Materialy, Trudy (Naukova Dumka, Kiev) 27:32
82. Batog AE, Savenko TV, Pet'ko IP (1980) Reaktsionnosposobnye Oligomery, Polimery i Materialy na Ikh Osnove (NPO "Plastik", Moscow) 3:37
83. Pet'ko IP et al. (1986) unpublished results
84. Pet'ko IP, Pandazi IF, Batog AE (1980) Plast Massy (2):28
85. Durmanenko NA et al. (1976) unpublished results
86. Pet'ko IP et al. (1974) Plast Massy (5):60
87. Pet'ko IP et al. (1986) unpublished results
88. USSR Pat 1,513,866
89. USSR Pat 1,623,147
90. USSR Pat 1,705,294
91. Batog AE, Sorokin VP (1984) Plast Massy (4):36

Remark:"Plast. Massy" means "Plasticheskiye Massy" (Moscow)

Editor: Prof. K. Dušek
Received: October 1998

Recent Advances in the Study of Synthetic Polyampholytes in Solutions

Sarkyt E. Kudaibergenov

Department of High Molecular Compounds, Faculty of Chemistry, Kazak State National University, Karasai Batyra Str.95, 480012, Almaty, Republic of Kazakhstan
e-mail: skudaibe@aad.alma-ata.su

This article reviews the theoretical and experimental material accumulated in recent years on water-soluble and water-swelling polyampholytes of integral and pendant types. The theory of polyampholytes in comparison with experimental results, acid-base equilibrium, hydrodynamic, conformational, and the molecular and colloidal properties of amphoteric copolymers are considered. The responsibility of Coulomb forces, hydrogen bonds and hydrophobic interactions in the formation of compact structures is shown. Basic attention is paid to the specific structure and behaviour of polyampholytes at or near the isoelectric point (IEP), in particular to the "forcing out" phenomenon found at the IEP. The ability of polyampholytes to associate with various low- and high-molecular-weight substances as well as to be absorbed on dispersed particles is discussed. An attempt to show the closeness of structural organization or molecular recognition of synthetic polyampholytes to natural ones is undertaken.

Keywords: Polyampholytes, Theory of Polyampholytes, Acid-Base Equilibrium, Solution Properties, Associates and Complexes, Zwitterions, Amphoteric Gels, Application

List of Symbols and Abbreviations

Sect. 3

γ	dimensionless parameter
e	electron charge
b	monomer size
l	Kuhn length
L	chain length
k	Boltzmann constant
T	absolute temperature
N	monomer number
f	fraction of positively charged monomer
g	fraction of negatively charged monomer
k^{-1}	Debye length
l_B	Bjerrum length
k_p^{-1}	screening length
Θ	Flory's Θ temperature
R	Einstein sphere radius
V	excluded volume
V^*	effective excluded volume
N_e	polyelectrolyte blob size
N_1	polyampholyte blob size
$\pm q$	excess of positive or negative charges
Q_c	overall excess charge
N_+	number of positively charged monomers
N_-	number of negatively charged monomers
t	effective temperature
u	ratio of Bjerrum length to bond size
D	electrostatic blob size
σ	charge asymmetry
r_D	Debye volume

Sect. 4

pK'_a	apparent acidic dissociation constant of acidic group
pK'_b	apparent acidic dissociation constant of basic group
α	ionization degree of acidic group
β	ionization degree of g basic group
$pK°$	intrinsic dissociation constant
E_d	nearest-neighbour electrostatic interaction
E_r	long-range electrostatic interaction
$pK°_a$	intrinsic dissociation constant of acidic group
$pK°_b$	intrinsic dissociation constant of basic group
ν	mean number of negative charge
ξ	mean number of positive charge
$|\psi_a|$	electrostatic potential
α_{IEP}	ionization degree of acidic group at the isoelectric point
β_{IEP}	ionization degree of basic group at the isoelectric point
a	molar fraction of acidic units
R	molar ratio of acid to base

Sect. 5

$[\eta]$	intrinsic viscosity
S_0	sedimentation constant
$M_{s\eta}$	average hydrodynamic molecular weight
μ	ionic strength of the solution
η_i/c	reduced viscosity
M_w	weight average molecular mass
R_g	gyration radius
R_h	hydrodynamic radius
C_p	polymer concentration
C_s	salt concentration
d_h	hydrodynamic diameter
d_g	gyration diameter
A_2	second virial coefficient
T_1	spin-lattice relaxation time
τ_c	correlation time

Sect. 6

σ	surface tension
χ	electroconductivity
C_{crit}	critical concentration
G	surface activity
Ω	fraction of precipitated polyampholyte
C_{PA}^d	concentration of polyampholyte in disperse phase
C_{PA}^s	concentration of polyampholyte in solution
r_λ	average radius of disperse particles
τ	turbidity

Sect. 7

M	metal ions
L	ligand
$k_1, k_2,$	
$k_3 \ldots$	constants of stepwise complex formation
β	stability constant of polymer-metal complexes
n	average number of ligands bound to metal ion
$[L]_t$	total concentration of ligands
$[M]_t$	total concentration of metal ions
$[L]$	concentration of free ligands
$[LH]$	concentration of protonated ligands
$g_\perp$	perpendicular g-factor
$g_\parallel$	parallel g-factor
$A_\parallel$	constant of superfine structure
C_{PEC}	concentration of polyelectrolyte complexes

Sect. 9

W	retardation factor
U	potential energy of interaction
R	distance between particles centres
A	optical density
ρ	density of latex particles
C_L	concentration of latex particles
B	optical constant
k_0	rate constant of fast coagulation
V_C	adsorption threshold
$\tilde{A}$	value of specific adsorption
$k_{1,2}$	adsorption constant rate
η	medium viscosity
r_1	average radius of latex particles
r_2	average radius of polyampholyte coils
t_A	adsorption time
f	effectiveness of adsorption
N_1	concentration of latex particles
ζ	electrokinetic potential

Sect. 10

S	sedimentation coefficient
I/I_0	relative fluorescence intensity
Ξ	degree of phase formation

Sect. 11

a	exponent of Mark–Kuhn–Houwink equation
α_e	expansion factor
T_g	glass transition temperature

| D | dipole moment |
| E_A | activation energy |

Sect. 12

V/V_0	volume swelling ratio
d	diameter of gel
d_0	initial diameter of gel

Abbreviations

IEP	isoelectric point
pH_{IEP}	pH of the isoelectric point
2VP–AA	2-vinylpyridine–acrylic acid
2VP–MAA	2-vinylpyridine–methacrylic acid
2M5VP–AA	2-methyl-5-vinylpyridine–acrylic acid
1-VI	1-vinylimidazole
DMAEM–MAA	N,N'-dimethylaminoethylmethacrylate-methacrylic acid
PDMAEM	poly(N,N-dimethylaminoethylmethacrylate)
DMAEM–MMA–MAA	N,N'-dimethylaminoethylmethacrylate-methylmethacrylate–methacrylic acid
CM-DEAEC	carboxymethyl-2-diethylaminoethylcellulose
TMVEP–MAA	1,2,5-trimethyl-4-vinylethynylpiperidol-methacrylic acid
CM-HTMAPC	O-carboxymethyl-O-2-hydroxy-3-(trimethylammonio)propylcellulose
2M5VP–MAA	2-methyl-5-vinylpyridine–methacrylic acid
AMPDAC–SAMPS	2-acrylamido-2-methylpropyldimethylammonium chloride–sodium2-acrylamido-2-methylpropanesulfonate
METMAC–SAMPS	2-methacryloyloxyethyltrimethylammonium chloride–sodium 2-acrylamido-2-methylpropanesulfonate
METMAC–AAm–SAMPS	2-methacryloyloxyethyltrimethylammonium chloride–acrylamide–sodium 2-acrylamido-2-methylpropanesulfonate
PSS-b-P2VP	poly(styrene sulfonate)-block-poly(2-vinylpyridine)
IPC	interpolyelectrolyte complex
PMAA-b-P1M4VPCl	poly(methacrylic acid)-block-poly(1-methyl-4-vinylpyridinium chloride)
CCME	chitosan carboxymethyl esters
BSA	bovine serum albumin
SAXS	small angle X-ray scattering
MAA–DMAMA	methacrylic acid–dimethylaminomethacrylate
NaMES–METMAI	sodium 2-methacryloyloxyethanesulfonate-2-methacryloyloxyethyltrimethylammonium iodide

DMVEP–AA	2,5-dimethyl-4-vinylethynylpiperidol–4-acrylic acid
PPG	poly(N-propyleneglycine)
PIPCEI	poly(1-isopropylcarboxylethyleneimine)
PEA	poly(ethylenealanine)
EDTA	ethylenediaminetetraacetic acid
VEEA–MAA	vinyl ether of ethanolamine–methacrylic acid
DMF	dimethylformamide
DDSNa	sodium dodecylsulfate
CTMACl	cetyltrimethylammonium chloride
MDAA–MA	N-methyldiallylamine–maleic acid
PDMDAAC	poly(N,N-dimethyldiallylammonium chloride)
DMDAA–MA	N,N-dimethyldiallylammonium–maleic acid
PAA	poly(acrylic acid)
PEC	polyelectrolyte complexes
PVBTMACl	poly(vinylbenzyltrimethylammonium chloride)
PVPD	poly(N-vinylpyrrolidone)
PEG	poly(ethyleneglycol)
C_{PEC}	concentration of PEC
PMAA	poly(methacrylic acid)
MVEP–MAA	1-methyl-4-vinylethynylpiperidol–4-methacrylic acid
m_{HSA}	mass fraction of human serum albumin
m_{PA}	mass fraction of polyampholyte
PSL	poly(styrene) latex
HSA	human serum albumin
IG	immunoglobulins
DLVO	Derjaguin–Landau–Verway–Oberbeek
PEI	poly(ethyleneimine)
ANS	1-anilino-8-naphthalenesulfonic acid
AY	acrylidine yellow
SPSS	sodium poly(styrene sulfonate)
DEAEM–MAA	N,N-diethylaminoethylmethacrylate–4-methacrylic acid
MKH	Mark–Kuhn–Houwink equation
PDMAAPS	poly[N,N-dimethyl(amidopropyl)ammonium propanesulfonate]
P2VP-SB	poly(2-vinylpyridinesulfonatopropyl betaine)
P4VP-SB	poly(4-vinylpyridinesulfonatopropyl betaine)
UCST	upper critical soluble temperature
LCST	lower critical soluble temperature
APTAC	acrylamidopropyltrimethylammonium chloride
SA	sodium acrylate
MAPTAC–AA	methacrylamidopropyltrimethylammonium chloride–acrylic acid
SSS	sodium styrene sulfonate

NIPA	N-isopropylacrylamide
DC	direct current
TRR	thermoregenerable resins

1
Introduction

In order to adapt the physico-chemical properties of living materials to their biological functions, nature has developed polyelectrolytic polymers with outstanding physical and chemical behaviour. Among the synthetic polyelectrolytes, polyampholytes are very close to biological macromolecules in nature and behaviour [1]. They may provide useful analogs of proteins and are important to modeling some properties and functions of biopolymers such as protein folding and enzymatic activity. The latter has been demonstrated recently [2] and is not considered here.

The dualistic character of polyampholyte chains as compact particles and expanded conformation, as well as the existence of isoelectric point where the net charge of the whole macromolecule is zero and which divides the state of polyampholytes into polyacids and polybases, is also exciting. In spite of our previous publications [3–5], the recent progress in studying the behaviour of synthetic polyampholytes, mainly due to the remarkable works of several research centres and groups on the theory (Edwards, Joanny, Kholodenko, Kardar, Dobrynin, Grosberg and Khokhlov) and the synthesis, characterization and aqueous solution properties of linear and cross-linked polyampholytes (Salamone, McCormick, Candau, Merle, Laschewsky, Tanaka, Katayama), invoked us to systematize and analyze the theoretical and experimental materials in order to predict the perspectives for developing such an interesting class of polyelectrolytes. One of the main purposes of this report is also to bridge the gap between synthetic and natural polymers, between biological materials and the physics and chemistry of macromolecules, and to show a closer relationship between two fascinating worlds.

2
Classification of Polyampholytes

Polyampholytes can comprise chains with combinations of weak acid/weak base, strong acid/weak base, (or else weak acid/strong base) and strong acid/strong base monomers. Typical polyampholytes consisting of weak base and weak acid groups are copolymers of vinylpyridines and acrylic (or methacrylic) acid. While copolymers of N-substituted allylamines and vinyl- or styrenesulfonic acids belong to strong base/strong acid polyampholytes, the acidic and basic groups of polyampholytes can also be in salt forms with a low or high charge density along the macromolecules. Independent of the microstructure, polyampholyte chains can be divided into a random (1), alternating (2), graft (3), diblock (4) or triblock (5) sequence.

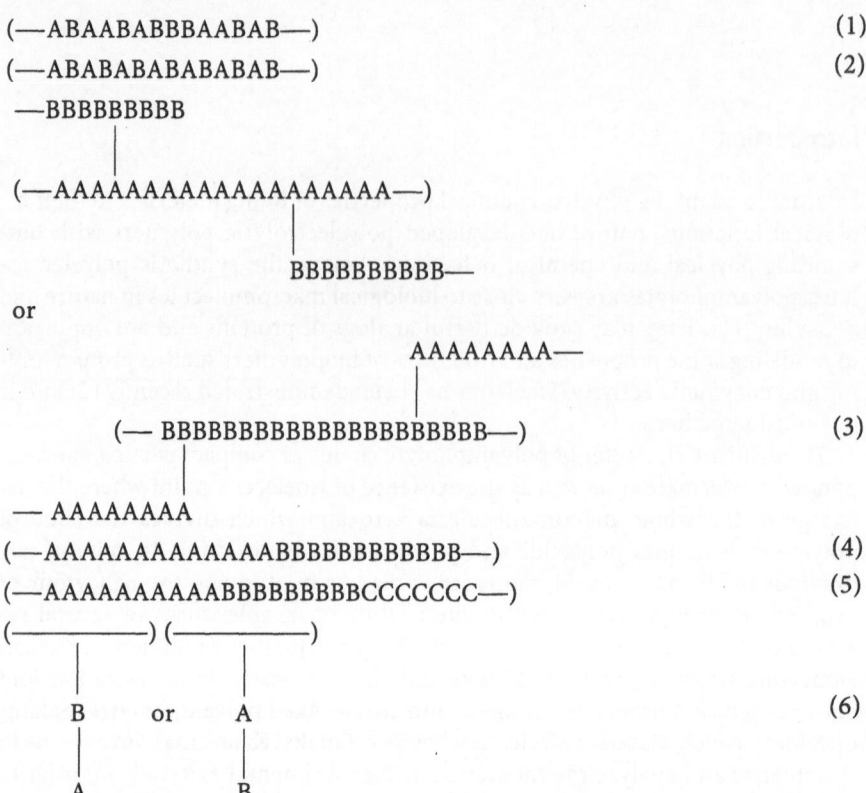

(—ABAABABBBAABAB—)   (1)

(—ABABABABABABAB—)   (2)

—BBBBBBBBB

(—AAAAAAAAAAAAAAAAAAAA—)

BBBBBBBBBB—

or

AAAAAAAA—

(— BBBBBBBBBBBBBBBBBBBBBB—)   (3)

— AAAAAAAA

(—AAAAAAAAAAAAABBBBBBBBBBBB—)   (4)

(—AAAAAAAAAABBBBBBBBCCCCCCC—)   (5)

(———————) (———————)

B   or   A   (6)

A   B

where A, B and C are acidic, basic and neutral monomers, respectively.

In the case of polyampholytes with a betaine structure (6), the acidic and basic groups are situated along the chain backbone. Dependent on the solubility of polyampholytes near the isoelectric point (IEP), they can be water soluble and water insoluble. For instance, the equimolar copolymers of aminoalkyl(meth) acrylates and unsaturated carboxylic acids are water soluble over the complete range of pH-values. On the contrary, copolymers based on vinylpyridines and acrylic (methacrylic, vinyl- or styrenesulfonic) acid are insoluble at the IEP. As a rule most blockpolyampholytes have a wide region of insolubility. Hydrophobic polyampholytes, the behaviour of which is close to "polysoaps", represent the combination of zwitterionic and hydrophobic structures.

3
Theory of Polyampholytes

The theory of solutions of flexible uncharged polymers with excluded volume is at present well developed, but the properties of polyelectrolytes and especially polyampholytes have been considered much less from the theoretical point of view. It is well known that polyampholytes exhibit a change in phase from the extended random flight configuration to a condensed microphase. The polyampholyte theory of Edwards et al. [6] considers the isoelectric state of polyampholytes as a microelectrolyte satisfying a Debye–Huckel-type of structure. The criterion of transition from the collapsed conformation to the extended one is described as follows:

$$\gamma = \frac{e^2 b^2}{kT}(Ll)^{-1/2} \gg 1 \quad \text{collapse}$$
$$\ll 1 \quad \text{open chain}$$

$$(3.1)$$

where γ is a dimensionless parameter; e is the electron charge; b is the monomer size; l is the Kuhn length; L is the arc length; k is the Boltzmann constant; and T is the absolute temperature.

The concluding remarks of this work are that the addition of salt to the solution will move the collapse point. In the case of charged blockpolyampholytes the existence of a nonuniform dumbbell-like configuration distributed in space is possible. One half of this dumbbell configuration will have a surplus positive and the other half a surplus negative charge.

Higgs and Joanny [7] considered the behaviour of positive and negative charges inside the globule to be similar to that of charges in a small volume of a simple electrolyte. The distribution of the charged monomers within the globule resembles that of the charges in a simple ionic solution rather than those in an ionic crystal. The electrostatic interactions create correlation between the positions in space of the positive and negative charges so that each charge tends to be surrounded by a shell of charge of the opposite sign. The Coulomb interactions are thus screened at long distance and the charges on the polyampholyte chain can be considered as free charges in an ionic solution. The Higgs and Joanny model considers a single chain having a fraction of positively and negatively charged monomers randomly distributed along the macromolecule. Let a chain of N monomers of size b have a fraction f of positively and a fraction g of negatively charged monomers distributed randomly. For a neutral chain ($f=g$) in pure water it is supposed that there exists a collapsed state of the chain with a volume significantly smaller than the Gaussian chain volume $N^{3/2}b^3$ but significantly larger than the volume of the completely close-packed monomers Nb^3. The chain is at a thermodynamic equilibrium state and can easily change from one configuration to another. If the chain is in a Θ solvent where the second virial coefficient is zero the third virial term will counteract the electrostatic attractions. If the polymer is an Einstein sphere of radius R, the charge concentration in this region is $\cong fN/R^3$. By analogy with the Debye–Huckel screening length k^{-1}, one can define a screening length k_p^{-1} due to the charge on the polymer to be given by:

$$k_p^2 \cong fNl_B/R^3 \tag{3.2}$$

where l_B is the Bjerrum length.

It is necessary to consider different regimes dependent on the dimensions of polymer chains. A long chain collapses into a globule consisting of close-packed blobs of size k^{-1}. The blobs are either Gaussian or swollen depending on the second virial coefficient value. In a Θ solvent, the chain radius in the collapsed state is given by:

$$R \cong N^{1/3} b (\frac{b}{fl_B})^{1/3} \tag{3.3}$$

In the case of a finite concentration of chains, Higgs and Joanny predict a phase separation between a high concentration phase of interpenetrating polymer chains and a dilute solution in which each chain is a collapsed globule. In the presence of salt, due to the screening of the electrostatic attractive forces by low-molecular-weight ions, the polyampholyte molecules tend to swell and adopt the dimensions of the uncharged chain. Then the electrostatic attractions are treated as a negative contribution to the excluded volume. Therefore a Θ compensation point exists at the salt concentration at which the positive excluded volume interaction V balances the negative electrostatic contribution. The effective excluded volume is given by:

$$V^* = V - 4\pi l_{B^2} f^2 / kb^3 \tag{3.4}$$

At the Θ point ($V^*=0$), the screening length is:

$$k = k_\theta = (l_B f / b)^2 (1/Vb) \tag{3.5}$$

This salt concentration marks the transition from collapsed to Gaussian or swollen chains.

Higgs and Joanny also considered a non-neutral polyampholyte with $f-g <<f+g$ for the case when $V=0$. In pure water the chain has the characteristics of both a polyelectrolyte (excess of charges: $f-g$) and of a neutral polyampholyte (neutral part: $2g \approx f+g$). The polyelectrolyte blob size is:

$$N_c \sim (\frac{b}{l(f-g)^2})^{2/3} \tag{3.6}$$

Hence, the polyampholyte blob size is:

$$N_l \sim \frac{b^2}{l^2(f+g)^2} \tag{3.7}$$

The overall conformation of the chain is determined by the relative value of N_1 and N_e. If $N_1 < N_e$ the attractive polyampholyte interaction dominates. The polymer collapses in the same way as a neutral polyampholyte. If $N_1 > N_e$ the repulsive polyelectrolyte interaction dominates and the chain behaves more like a usual polyelectrolyte. The necessary net charge to swell, and thus redissolve the polyampholyte, is therefore:

$$f - g > l_B / b(f + g)^{3/2} \tag{3.8}$$

At a finite salt concentration the polyampholyte effect dominates the polyelectrolyte effect at long distances if:

$$(f - g)^2 < \frac{k l_B}{4}(f + g)^2 \tag{3.9}$$

Hence the chain is expected to collapse for a wider range of $(f-g)$ at higher salt concentrations. For a small net charge case, the reverse of Eq. 3.8, the polyampholyte term dominates at small distances. The main difference between the polyampholyte theory of Edwards et al. [6] and that of Higgs and Joanny [7] is that the former argues that the electrostatic attractions are balanced by the reduction in entropy of the chain due to confinement within a sphere smaller than its natural radius while the latter considers that the third order virial term turns out to be always larger than the confinement entropy for a collapsed chain and must be taken into account. Equation 3.1 also seems to state that short chains collapse and long ones are open which is contrary to that found by Higgs and Joanny. The original Debye–Huckel theory is only valid at low salt concentrations $k l_B < 1$, and, therefore, Higgs and Joanny's theory is not quantitatively accurate for the high salt concentrations needed to dissolve the highly charged polyampholytes.

Two recent studies [8,9] have led to opposite conclusions as to whether the configuration of a polyampholyte is stretched [8,9] or compact [7]. The disagreement between these two conclusions is quite extreme, in that one predicts a compact configuration employing the Debye–Huckel theory, while the other substitutes dimensional arguments for a true renormalization-group treatment. The Debye–Huckel approach implicitly assumes that the polymer is overall neutral (or at least very close to neutrality), while the scaling results rely on typical excess charges in the order of $\pm q_0 N^{1/2}$ in the various quenched configurations. It has been proved that the conformation of the polyampholyte is very sensitive to the presence of constraints and the choice of ensemble. A polymer constrained to have a zero net charge collapses to a compact configuration upon reduction in temperature. On the other hand, if N monomers are randomly assigned positive or negative charges, $\pm q_0$ with equal probability, the resulting polymer has an overall excess charge $Q_c \approx \pm q_0 N^{1/2}$. Independent of its sign this deviation from neutrality is sufficient to stretch the chain to an extended state at low temperature. Indeed the neutral polyampholyte collapses at low temperatures as predicted by Higgs and Joanny [7]. This result is not particularly surprising for relatively short chains with a moderate excluded-volume interaction, due to the crossover between Gaussian and self-avoiding behaviours. Very different results are obtained when the charges are selected randomly without regard to overall neutrality. As the temperature decreases below T_0, the radii first contract and then reexpand. The explanation of this phenomenon is as follows: at T somewhat smaller than T_0, positive and negative charges start to pair up, thereby reducing the chain size. At lower T the "excess charges" can only reduce their energy by stretching the entire chain, the chain "unfolds" into a stretched linear or branched object with folded "double strands".

In a very recent work, Gutin and Shakhnovich [10] considered the effect of a net charge on chain conformation. They introduced an interesting idea of the elongation of a polyampholyte globule. Their calculations, however, for asymmetric polyampholytes, led to the wrong temperature dependence of the aspect ratio of a polyampholyte chain. According to Dobrynin and Rubinstein [11], the net charge on a typical polyampholyte chain is hardly ever exactly zero because it is usually made by random copolymerization and therefore has some small net charge even if the whole system of chains is, on average, neutral. In this connection the conformation of a single polyampholyte chain was reexamined in the framework of a simple two parametric Flory theory. Let us consider N-mer with $N_+=fN$ positive and $N-=gN$ negative charges. There are three different regimes for polyampholyte chains with charge asymmetry $|N_+-N_-| > |N_+-N_-|^{1/2}$ depending on the effective temperature $t=N/u(N_+-N_-)$ (where u is the ratio of Bjerrum length to the bond size) in a salt-free solution:

i) *Unperturbed regime*: at high effective temperatures $t > t_1$ the electrostatic interactions are unimportant and polyampholyte configuration is controlled by the solvent quality for the uncharged backbone. It is swollen in good solvent, Gaussian in the Θ solvent, or collapsed in a poor solvent (Fig. 1a);

ii) *Polyelectrolyte regime*: at intermediate effective temperatures $t_2 < t < t_1$ the repulsion between uncompensated charges $e(N_+-N_-)$ dominates over the charge-fluctuation-induced attraction. The chain is stretched into an array of electrostatic blobs each of size D (Fig. 1b). Inside these blobs the chain statistics are unperturbed by the electrostatic interactions and are the same as in Regime (i). On length scales larger than D the chain is strongly stretched into an array of blobs of total length L. The aspect ratio of L/D of a polyampholyte chain is equal to the number of electrostatic blobs. This aspect ratio is proportional to N and increases with an increase in charge asymmetry $(f-g)$ and with a decrease in temperature T. It should be stressed that the repulsion between uncompensated charges cannot be screened by other charges on the chain.

iii) *Polyampholyte regime* $(t < t_1)$: in this regime the aspect ratio of a polyampholyte chain is constant and equal to its maximum value:

$$L/D \approx (N_+ - N_-)^2 / (N_+ - N_-) \text{ for } t \leq t_2 \qquad (3.10)$$

Fluctuation-induced attraction reduces the chain size (reduces the volume at constant aspect ratio) with decreasing temperature. On length scales smaller than the Debye radius r_D the polymer configuration is unperturbed [the same as in Regime (i)]. On larger length scales the polyampholyte is a dense packing of Debye blobs which form an asymmetric globule as sketched in Fig. 1c. It is interesting to note that the length L of the polyampholyte globule could be either larger or smaller than the unperturbed size of the chain depending on the reduced temperature range. Chains with smaller charge asymmetries $(N_+-N_-)^2/(N_+-N_-)<1$ do not exhibit Regime (ii) (i.e. they do not stretch beyond their unperturbed size) and directly collapse into a globular state [Regime (iii)] from the unperturbed state [Regime (i)] with decreasing temperature. These "almost neutral" polyampholytes form symmetric globules (with aspect ratio $L/D≈1$). The shape anisotropy of a globule, predicted for

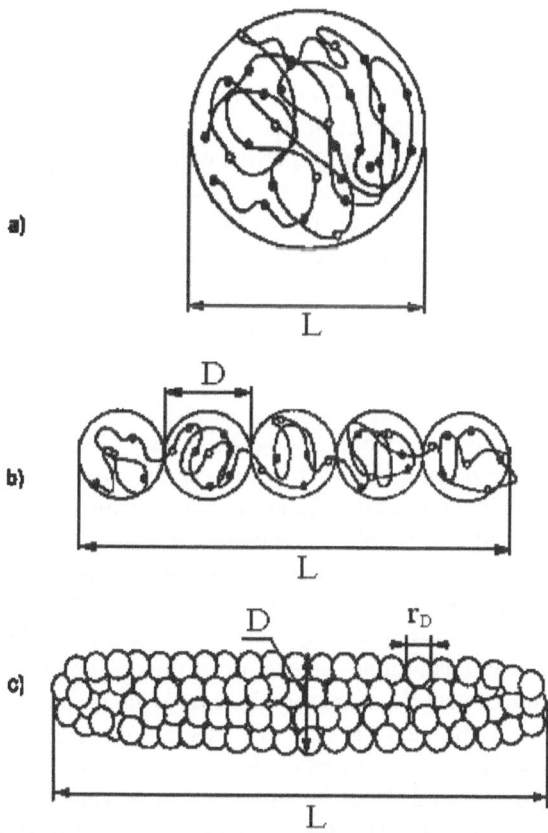

Fig. 1a-c. The three regimes of a polyampholyte chain with charge asymmetry $(N_+-N_-)^2>$ (N_++N_-): **a**) unperturbed regime $(t>t_1)$; **b** polyelectrolyte regime $(t_2<t<t_1)$; **c**) polyampholyte regime $(t>t_2)$ – an elongated globule of Debye blobs (represented by *circles*) [11]

the polyampholyte regime, could be measured in super-cooled polyampholyte solutions. It should also have a profound effect on the polyampholyte phase diagram.

The predictions of references [7] for the case $N_+=N_-$ coincide well with the conclusions of authors [11]. For asymmetric polyampholytes $N_+\neq N_-$, Higgs and Joanny [7] predict that electrostatic repulsion swells the chain only for very large charge asymmetry $\sigma N >t^{-2}$, whereas Dobrynin and Rubinstein [11] found that electrostatic repulsion can stretch the polyampholyte chain with much smaller charge asymmetry $\sigma N >1$ (of the order of statistical charge fluctuation of the number of charges on the chain).

Qian and Kholodenko [12] studied the conformational properties of polyampholytes according to a Bernoullian type of statistics. It has been shown that the mechanism of collapse transition for the random polyelectrolytes essentially differs from that of uncharged homopolymers. Below the collapse transition temperature, depending on the external conditions, polymer chains can be in one of two globular phases with the possibility of a quasi-first order (due to finite size effects) transition between them. This fact has important biological significance with regard to enzyme-substrate complexes.

The dynamics and conformational properties of polyampholytes have been studied in the presence of external electric fields. In terms of the Rouse model of polymer dynamics, some authors [13] obtained the main-square displacement both of the centre of mass and of individual beads, and the end-to-end distance of the polyampholyte. The equilibrium mean-square end-to-end distance shows an additional elongation which is represented as N^β: the exponential equals $\beta=2$ if only one bead is charged or if the charges are distributed in an alternating fashion, $\beta=3$ if the charges are distributed in an uncorrelated fashion, and $\beta=4$ if the correlation length of the charge distribution is large.

Muthukumar and co-workers [14] have studied 20 different randomly charged polyampholytes of length 50 using Monte Carlo simulations. Results for six representative sequences from among the twenty sequences are presented. The mean-squared radii of gyration R_g^2 of the chains are a function of temperature for these sequences. At high temperatures the chains behave as self-avoiding walks and the radii are similar. As the temperature is reduced the radii decrease under the influence of attractive electrostatic interactions. The distribution of the radii and of intrachain energy at three different temperatures for two sequences has been simulated. The location of charges along the chain plays a crucial role in the conformational behaviour of the polyampholytes. In the authors' opinion, the knowledge of merely the energy landscape is insufficient to describe the conformational properties of polyampholytes. The entropy associated with the formation of frustrated conformations needs to be accounted for when trying to understand the conformational properties of heteropolymers with specific sequences. Monte Carlo simulation and variational mean field calculations were used to study the structure of isolated polyampholyte chains at conditions roughly corresponding to dilute solutions [15]. A random distribution of cationic and anionic groups on the chain is assumed and average properties for samples with restricted or fluctuating net charge on individual chains are computed. The chains swell with increasing net charge while they contract when a balance of positive and negative charges is attained. The variational mean field theory successfully describes the swelling at high net charge while it underestimates the attractive effects characteristic of neutral or nearly neutral chains. This difference is interpreted as the result of spatial correlations among ionized polyampholyte beads in compact coils.

All the above-mentioned theoretical conclusions successfully explain most experimental results as follows:

1) When $f \gg g$ or $g \gg f$, polyampholyte chains are close to a polycation or a polyanion, respectively, and behave as polyelectrolytes.
2) Polyampholytes at the isoelectric state when the whole macromolecule is neutral tend to collapse, have a small hydrodynamic radius if soluble, and have the phase separation if insoluble.
3) The addition of simple salts leads to swelling of neutral polyampholytes if they are soluble and dissolve the precipitate if they are insoluble ("antipolyelectrolyte" effect).

There are also some discrepancies between the theory and the experimental. All the above-mentioned theories developed for polyampholytes:

1) Neglect the effect of counterions on the conformation of charged chains;
2) are applicable only at low concentration of salts;
3) consider only the diluted solutions;
4) ignore the specific binding of counterions by polyion.

The validity of theories at low salt concentrations is connected with the fact that the Debye–Huckel length k^{-1} is large with respect to the Bjerrum length l_B. The theory cannot be quantitatively accurate for the high salt concentrations needed to dissolve the highly charged polyampholytes, especially blockpolyampholytes. Usually the hydrodynamic properties of amphoteric copolymers are studied in the dilute regime, while the solubility data refer to the semidilute range; NMR or Raman spectroscopic experiments are sometimes performed in the concentrated regime. The preferential binding of anions by polyampholytes usually leads to the shifting of pH_{IEP} to the acidic region or, vice versa, the specific binding of cations shifts the pH_{IEP} towards the alkaline region [16]. The effectiveness of anions to decrease the pH_{IEP} changes mostly as follows: ClO_4^- $>SCN^- >NO_3^- >I^- >Cl^-$ and coincides well with the Hoffmeister series for proteins. In spite of these reservations, the above-mentioned models describe reasonably well the solution behaviour of most polyampholytes.

4
Acid-Base Equilibrium in Polyampholytes

Both the nature of their functional groups and the microstructure of the chain determine the electrochemical properties of polyampholytes. The acid-base properties of amphoteric polyelectrolytes are quite different from the behaviour of both polyacid and polybase due to the fact that near the IEP it is difficult to titrate the functional groups of polyampholytes [1]. The titration of acidic and basic groups becomes available away from the IEP when positive or negative charges begin to predominate. Then a modification of the Henderson–Hassel-bach equation can be used to analyze the acid-base equilibrium of polyampholytes [17–19]:

$$pH = pK_a^{'} + \log{(\alpha / 1 - \alpha)} \tag{4.1}$$

$$pH = pK_b^{'} + \log{(1 - \beta / \beta)} \tag{4.2}$$

where $pK_a^{'}$ and $pK_b^{'}$ are the apparent acidic dissociation constants of acidic and basic groups, respectively, and α and β are the degree of ionization of acidic and basic groups of polyampholytes, respectively.

To determine the dissociation constants of ionizing groups, Katchalsky and Gillis [20] have suggested a theoretical equation based on the model of electrostatic potential smeared along the chain molecule. It takes into account the electrostatic interaction between different chain segments as well as between ionic groups and low molecular weight electrolytes. The theoretical calculations are in good agreement with experimental results obtained for 2-vinylpyridine–acrylic acid (2VP–AA). Analysis of the titration curves of amphoteric copolymers of 2-vinylpyridine–methacrylic acid (2VP–MAA) [21] and 2-methyl-5-vinylpyridine–acrylic acid (2M5VP–AA) [22, 23] of different composition reveals that at low MAA (or AA) content they show a dependence analogous to that of polyvinylpyridines alone but, upon increasing the acid content, the shape of the titration curve changes. The dissociation constant for each copolymer sample depends on its composition: increasing the MAA (or AA) content enhances acid ionization of pyridine groups, while increasing the 2VP content decreases it. The increasing of the 1-vinylimidazole (1-VI) content from 24 to 56,4 mol% enhances the acidity of the AA group's pK_a from 3.97 to 3.25, whereas increasing the AA content from 43.6 to 74 mol% enhances the basicity of 1-VI's pK_b from 10 to 11.4 [4]. This behaviour by polyampholytes is explained by the inductive influence of neighbouring groups. Rice and Harris [24,25] have extended the chain model suggested for polyelectrolytes containing charges of the same sign to alternating polyampholytes. The model takes account of the equilibrium dissociation of acidic and basic groups, counterionic binding, as well as the electrostatic interaction between chain charges. The electrostatic free energy of interaction between nearest-neighbour charges is calculated by the iteration method. Calculations of potentiometric titration curves for a hypothetical copolymer show that the properties of polyampholytes qualitatively approach those of proteins. Katchalsky et al. [17] considered the case when one monomer is present in the copolymer in such abundance that formation of isolated groups in the polymer chain occurs. The results of the theoretical study are compared with the experimental data for 2VP–AA (MAA) synthetic copolymers and for the natural polyampholyte–lysozyme. The agreement between theory and experimental is satisfactory. To extend the utility of the Kachalsky model [17], the nearest-neighbour interaction for a polyampholyte of any composition was computed [19]. The calculated titration curves were compared with the experimental ones. The agreement between them is satisfactory. Ising models have been used to describe the acid-base behaviour of linear polyampholytes and polyelectrolytes [25]. Alternatively, such titration curves are often interpreted in terms of an affinity distribution (pK spectrum). The exact affinity distributions for various Ising models of linear polyelectrolytes and polyampholytes were derived. The acid-base equilibrium of polyampholytes depends also on the microstructure of copolymers. The solubility and titration curves, as well as pH_{IEP} are different for DMAEM–MAA prepared by radical initiated copolymerization of two monomers (A-type), alkaline hydrolysis (B-type) or acidolysis of poly-N,N-dimethylaminoethylmethacrylate (PDMAEM) (C-type), although the composition of the copolymers is the same. This is due to different distributions of the sequences of the acidic and basic monomers along the chain. Polyampholytes prepared by the hydrolysis of PDMAEM in concentrated sulfuric acid have a blocklike structure whereas an alkaline-hydrolyzed one tends towards randomness. The apparent pK's of the acidic and basic groups can be expressed by the sum of the intrinsic pK (pK°), the nearest-neighbour interaction (E_d), and the long-range electrostatic interaction (E_r):

$$pK'_a = pK°_a + 0.434 / kT \, (\partial E_d / \partial v)_\xi + 0.434 / kT \, (\partial E_r / \partial v)_\xi \qquad (4.3)$$

$$pK'_b = pK°_b + 0.434 / kT \, (\partial E_d / \partial \xi)_v + 0.434 / kT \, (\partial E_r / \partial \xi)_v \qquad (4.4)$$

where v and ξ are, respectively, the mean number of negative and positive charges on the macromolecule, k is Boatsman's constant, and T is the absolute temperature.

According to Katchalsky et al. [17]:

$$(\partial E_r / \partial v)_\xi = -(\partial E_r / \partial \xi)_v = e|\psi_a| \qquad (4.5)$$

where $e|\psi_a|$ is the potential on the surface of the polyion. At low ionic strength, the potential $e|\psi_a|$ is the sum of short- and long-range interactions; at high ionic strength only short-range interactions remain. If $pK°_a$ is the sum of the intrinsic pK (pK°$_a$) and the nearest-neighbour interaction term of the acid, and $pK°_b$ is the difference of the same terms of the base, we have:

$$pK'_a = pK°_a + 0.434 \, e|\psi_a|/kT \qquad (4.6)$$

$$pK'_b = pK°_b + 0.434 \, e|\psi_a|/kT \qquad (4.7)$$

Then the modified Henderson–Hasselbach Eqs. 4.1 and 4.2 can be rewritten as follows:

$$pH = pK°_a + \log(\alpha / 1 - \alpha) + 0.434 \, e|\psi_a|/kT \qquad (4.8)$$

$$pH = pK°_b + \log(1 - \beta / \beta) + 0.434 \, e|\psi_a|/kT \qquad (4.9)$$

The terms $pK^{\circ'}_a$ and $pK^{\circ'}_b$ can be obtained from Eqs. 4.8 and 4.9 if the neutralization is made at high ionic strength. At the IEP when the net charge is zero, $e|\psi_a|$ vanishes and Eqs. 4.8 and 4.9 become:

$$pH_{IEP} = pK^{\circ'}_a + \log(\alpha_{IEP} / 1 - \alpha_{IEP}) \tag{4.10}$$

$$pH_{IEP} = pK^{\circ'}_b + \log(1 - \beta_{IEP} / \beta_{IEP}) \tag{4.11}$$

If $pK^{\circ'}_b - pK^{\circ'}_a > 2$, the pH_{IEP} of a polyampholyte which contains an excess of acid monomer (a >0.5) is calculated from:

$$pH_{IEP} = pK^{\circ'}_a + \log(1 - a / 2a - 1) \tag{4.12}$$

and at an excess of base monomer (a <0.5) from:

$$pH_{IEP} = pK^{\circ'}_b + \log(1 - 2a / a) \tag{4.13}$$

Taking into account that $pK^{\circ'}_a = 5.15$ and $pK^{\circ'}_b = 8.58$ for MAA and N,N-dimethylaminoethylmethacrylate (DMAEM), respectively, an S-shaped curve of pH_{IEP} vs. the acid content has been obtained theoretically. The theoretical curve fits well with the experimental one for A-type copolymers. Small deviations from the theoretical curve are observed for copolymers of type **B** and **C** and can be explained by variations of $pK^{\circ'}_a$ and $pK^{\circ'}_b$ due to the nearest-neighbour interaction which depends on the distribution of the acidic and basic monomers. The values of pH_{IEP} of the DMAEM–MMA–MAA triblock copolymer [26] calculated from Eq. 4.14 form the theoretical curves of Fig. 2:

$$pH_{IEP} = pK_b + \log\{1 / 2[(1 - R / R) + (1 - R / R)^2 + (4 / R) \, 10^{pK_a - pK_b} \tag{4.14}$$

where R is the molar ratio of acid to base.

Similar to a simple amino acid – glycine, a polyampholyte dependent on pH, can change from the cationic to the anionic form, or vice versa, either through the molecular or the bipolar form. Apparently, the existence of full zwitterionic or molecular forms is not realized. The coexistence of both molecular and zwitterionic states is more probable [27]. Combination of potentiometric and conductimetric methods allowed for polyampholytes 2VP–AA and DMAEM–MAA to distinguish between: a) free COOH groups and $>NH^+$ – cations in the alkali titration of copolymers which have the acidic monomer in excess; b) the COO^- anions and >N-groups in the acid titration of copolymers which have an excess of basic monomer; c) COO^- anions of different basic strengths in the acid titration of a copolymer which has the acidic monomer in excess; d) >N-groups of different basic strengths in the acid titration of copolymers which have excess basic monomer [28,29]. The modified titration curves, $pH + \log(\beta / 1 - \beta)$ vs. β (where β is the degree of ionization of basic groups), for the copolymers show one or two discontinuities which correspond to several extrapolated values of pK°.

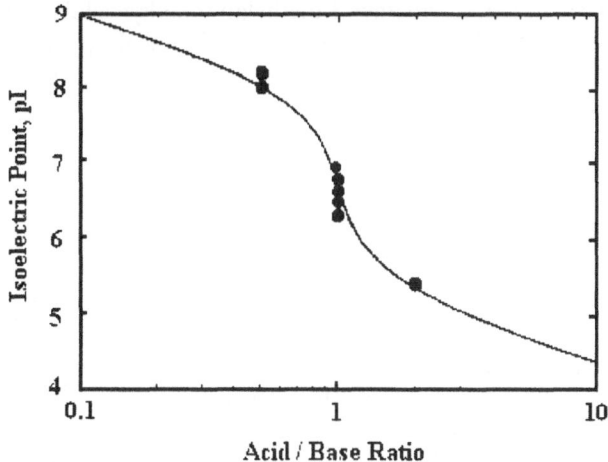

Fig. 2. Theoretical (*solid line*) and experimental (*filled circles*) values of the isoelectric points of DMAEM–MMA–MAA triblock copolymers (Reproduced with permission from [26])

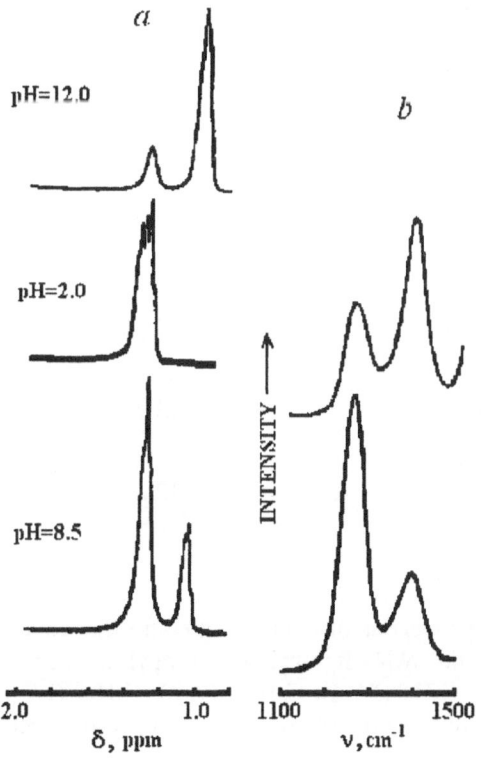

Fig. 3a,b. ^{1}H NMR **a** and FTIR **b** spectra of CM-DEAEC at various pH of the solution [30]

The existence of carboxymethyl-2-diethylaminoethylcellulose (CM-DEAEC) in protonated and non-protonated forms at different pH can be seen from ^{1}H NMR and FTIR spectra [30]. For methyl protons of DEAEC substituents only one signal is observed at 1.3 ppm in the acidic region whereas another strong signal at 1.05 ppm appears in the alkaline region (Fig. 3a). The former and the latter can be assigned to methyl protons of DEAEC substituents in protonated and non-protonated states. These spectra also indicate that about 80% of DE-AEC fragments exist in protonated form at pH=8.5. In spite of the high value pH of 12 some DEAEC groups still exist as cations. The FTIR spectrum of CM-DE-AEC has two peaks at v=1602 and 1740 cm^{-1} that correspond to C=O stretching vibrations of COO$^-$ and COOH, respectively (Fig. 3b). It can be concluded from FTIR measurements that the protonated DEAEC substituents promote the ionization of carboxylic groups through the transfer of protons from carboxylic to amine groups as mentioned earlier.

It has been shown [31] that the presence of hydrophobic groups in polyampholytes can change the equilibrium between the zwitterionic and molecular forms. Alternating copolymers of vinylethyl ether and N,N-diethylaminopropyl-monoamide of maleic acid preferentially exist in the charged form (I) but the replacement of ethyl radicals by two bulky butyl radicals stabilizes the uncharged molecular form (II).

$$
\begin{array}{ccc}
OC_2H_5 & & COO^- \\
| & & | \\
-(CH_2-CH)\!\!-\!\!\!-\!\!\!-(CH\!\!-\!\!\!-\!\!\!-CH)- & & \\
& | & \\
& CONH(CH_2)_3N^+HR_2 &
\end{array} \qquad (I)
$$

$$
\begin{array}{ccc}
OC_2H_5 & & COOH \\
| & & | \\
-(CH_2-CH)\!\!-\!\!\!-\!\!\!-(CH\!\!-\!\!\!-\!\!\!-CH)- & & \\
& | & \\
& CONH(CH_2)_3NR_2 &
\end{array} \qquad (II)
$$

where R=C$_2$H$_5$ (I) and C$_4$H$_9$ (II).

The acid-base properties of alternating comb-like copolymers of α-olefins (C$_6$H$_{12}$–C$_{16}$H$_{28}$) and N,N-dimethylaminopropylmonoamide of maleic acid [32,33] [where R=C$_4$H$_9$ (PA-4), C$_6$H$_{13}$ (PA-6), C$_8$H$_{17}$ (PA-8), C$_{10}$H$_{21}$ (PA-10), C$_{12}$H$_{25}$ (PA-12) with n=116, 59, 40, 21 and 16, respectively], are changed as shown in Table 1.

Table 1. Acid-Base Characteristics of Comblike Polyampholytes

Polyampholyte	n	pK_a	pK_b	pH_{IEP}
PA-4	116	3.62	10.82	7.22
PA-8	40	3.79	11.23	7.51
PA-10	21	4.24	11.28	7.71
PA-12	16	4.33	11.31	7,82

5
Hydrodynamic, Conformational and Molecular Characteristics of Amphoteric Copolymers

The copolymer composition, charge distribution and density, molecular weight and microstructure are the main parameters for a comprehensive analysis of the structure-properties-function relationships [4,5]. The hydrodynamic properties of polyampholytes are closely related to their conformations that are determined by the intra- or intermolecular interactions between anionic and cationic groups along polymer chains. An intramolecular attraction causes polymer chains to collapse into a globular structure, while an intermolecular one results in an association which behaves like colloid dispersions. A change in the hydrodynamic properties of 1,2,5-trimethyl-4-vinylethynylpiperidinol–4-methacrylic acid (TMVEP–MAA) as a function of pH medium is clearly seen from the data in Table 2 [34–36]. At the IEP the intrinsic viscosity [η] of TMVEP–MAA is minimal whereas the sedimentation constant S_0 has a maximal value. The closely spaced values of the average hydrodynamic molecular weights $M_{s\eta}$ at pH from 4.0 to 11.0 indicate the absence of intermacromolecular association. Figure 4 represents the influence of μ on the hydrodynamic sizes of O-carboxymethyl-O-2-hydroxy-3-(trimethylammonio)propylcellulose (CM-HTMAPC) [37]. In acidic and alkaline media, polyampholyte molecules behave as polycations and polyanions, respectively; the reduced viscosity value is decreased with the increase in μ. An "antipolyelectrolyte" effect is observed at the IEP: the reduced viscosity η_i/C increases with the increase in μ. These data, together with our previous results [34,35], entirely confirm the prediction of Tanford [1] and theoretical conclusions [7] on the unfolding of water-soluble amphoteric macromolecules at the IEP in the presence of neutral salts.

Table 3 lists hydrodynamic parameters of 2M5VP–MAA copolymers of different composition [38]. At a 90 mol% content of 2M5VP units at pH=1.2 the optical anisotropy of the polyampholyte is negative in sign and approaches the poly(2-methyl-5-vinylpyridine) anisotropy value. A change of the anisotropy sign from negative to positive is observed in the alkaline region. This indicates the formation of compact particles at pH=13, stabilized by the hydrophobic interactions of methyvinylpyridine groups composing the nuclei of the particles. In this case the polyampholyte macromolecule is protected from the precipitation due to hydrophilic carboxylic groups associated with water molecules.

Table 2. Hydrodynamic Characteristics of TMVEP–MAA (54:46 mol-%) at Various pH-Values and Ionic Strength of the Solution $\mu = 0.1$ mol·L^{-1}

pH	$[\eta]$, (dL·g^{-1})	$S_0 \cdot 10^{13}$ (s)	$M_{s\eta} \cdot 10^{-3}$
4.00	0.40	3.30	90.5
7.25	0.28	4.00	85.6
9.17 (IEP)	0.15	5.88	104.0
11.00	0.26	4.27	92,0

Table 3. Intrinsic Viscosities $[\eta]$, Weight Average Molecular Weight M_w and Gyration Radius R_g Values of Copolymers 2M5VP–MAA at Various pH-Values of the Solution

[2M5VP] (mol%)	pH	$[\eta]$, (g·dL^{-1})	$M_w \cdot 10^6$	R_g, (A)
90	1.2±0.05	5.00	0.90	700
	13.0±0.1	0.12	1.00	100
50	1.2±0.10	4.80	0.60	500
	13.0±0.1	4.00	0.60	500

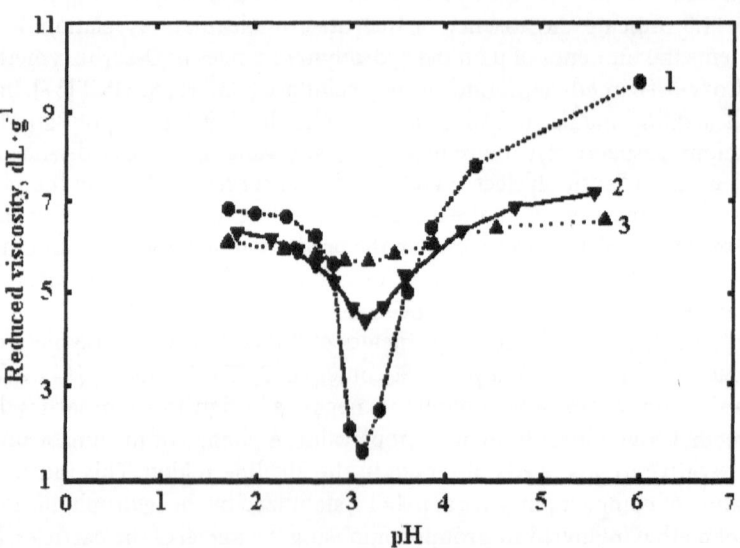

Fig. 4. Reduced viscosities of CM-HTMAPC (C_P=0.060 g·dL^{-1}) at different pH and ionic strengths of the solution μ. μ=0.01 (*curve 1*), 0.03, (*curve 2*) and 0.10 M NaCl (*curve 3*) (Reproduced with permission from [37])

Copolymers of 2-acrylamide-2-methylpropyldimethylammonium chloride and sodium 2-acrylamide-2-methylpropanesulfonate (AMPDAC–SAMPS) were prepared and their dilute solution properties studied as a function of copolymer composition, temperature, time, pH and added low-molecular-weight electrolytes [39–42]. Equimolar copolymers display a minimal viscosity in pure water and a maximal one in concentrated salt solutions. As the copolymer compositions deviate from an equimolar, charge imbalances increase the hydrodynamic volume of the macromolecules.

Corpart and Candau described the synthesis [43], characterization [44] and aqueous salt solution properties [45] of high charge density polyampholytes derived from 2-methacryloyloxyethyltrimethylammonium chloride–sodium 2-acrylamido-2-methylpropanesulfonate (METMAC–SAMPS) monomers. The polyampholytes prepared by means of a microemulsion polymerization have several advantages over other conventional procedures, namely, the process allows the production of very high-molecular-weight polymers (up to 1.7×10^7) and copolymers are more homogeneous in composition, with a microstructure not far from random. Solubility and viscometry experiments show that the solution behaviour of METMAC–SAMPS is essentially controlled by two parameters: the net electrical charge, fixed by the copolymer composition, and the ionic strength. Copolymers with a strong net charge behave more like polyelectrolytes. At intermediate copolymer compositions, the sample behaviour depends on the duality between repulsive (polyelectrolyte effect) and attractive (polyampholyte effect) electrostatic interactions. Depending on the ionic strength, either one of these antagonist effects will dominate. The same authors also reported the study of a low charge density polyampholyte METMAC–SAMPS with an incorporated acrylamide (Aam) monomer [46]. The influence of added salt on the solubility, as well as on the conformational and morphological characteristics of the terpolymer, was analyzed. The hydrodynamic parameters of METMAC–Aam–SAMPS (3.99:92.18:3.88 mol%) in relation to the ionic strength μ are summarized in Table 4. Figure 5 shows the schematic phase diagram of METMAC–AAm–SAMPS at two conditions: 1) when polymer concentration (C_P) is fixed and the salt concentration (C_S) varies; and 2) when the salt concentration is constant and the polymer concentration changes. At low salt content, attractive electrostatic interactions are dominant and the chains collapse. At high salt content, when the Debye–Huckel length becomes short enough so that the Coulomb interactions are screened out, then the excluded volume interactions are dominant and the polymer has a swollen or Gaussian conformation,

Table 4. Weight Average Molecular Weight M_w, Radius of Gyration R_G and Hydrodynamic Radius R_H of METMAC–AAm–SAMPS

μ (mol·L^{-1})	$M_w \cdot 10^{-6}$	R_G (Å)	R_H (Å)
0.2	9.5±1	2400±200	1500±100
0.5	9.1±1	2500±200	1500±100

depending on the quality of the solvent. Below a critical salt content, the chains with zero or small net charge precipitate, due to the polyampholyte effect. The supernatant contains highly swollen oppositely charged molecules. The polymer concentration in the supernatant decreases upon decreasing the salt content, the chains being more and more swollen. In other words, the phase separation observed here represents a fractionation process, not in terms of molecular weight distribution but rather in terms of a net charge distribution. Note that the mass polydispersity can also play a considerable role since the collapse–swollen transition is very sensitive to the molecular weight of the polyampholyte [7]. Various authors [47] have discussed multichain effects in polyampholyte solutions of finite concentration and found that the existing single-chain theories are limited to exponentially small concentrations if the sample contains net charges of both signs. It has been shown that the charged polyampholytes have a strong tendency to form neutral complexes and to precipitate. Stretched chains play almost no role in the ideal case of a neutral sample with nearly symmetric charge; counterions accumulate in the supernatant together with charged, elongated chains and the behaviour of the solution is dominated by "impurities".

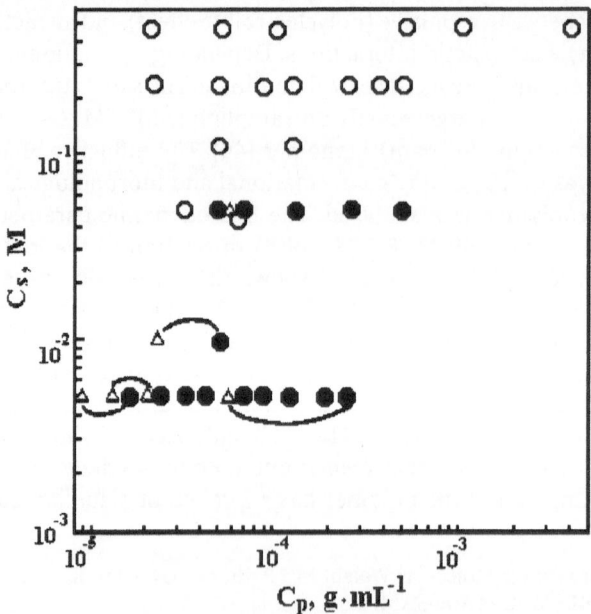

Fig. 5. Actual phase diagram for SAMPS/AAm/METMAC (3.83:92.18:3.99 mol%) copolymer. The *open circles* correspond to soluble samples, and the *filled circles* to samples for which a precipitate could be observed by eye; the *triangles* represent the concentrations of the supernatants estimated from the scattering intensities. The lines connect the initial systems to the corresponding supernatants (Reproduced with permission from [46])

Blockpolyampholytes possess unique physico-chemical properties owing to the combination of acidic and basic blocks in one macromolecular chain. It has been shown [48] that the properties of a linear diblock copolymer poly(styrenesulfonate)-*block*-poly(2-vinylpyridine) (PSS-*b*-P2VP) at the IEP are closely related to those of stoichiometric interpolyelectrolyte complexes (IPC). Depending on the degree of ionization of vinylpyridine blocks, the conformation of the macromolecules is changed as follows. In the uncharged state it is a segregated structure with a nucleus of vinylpyridine blocks maintained in water by hydrating polyacid blocks. At a low degree of ionization it is a coil "frozen" by rare ionic cross-links, at IEP it is a compact structure of IPC stabilized by the intraionic interactions of acidic and basic block units. The addition of a 1:1 type low-molecular-weight electrolyte leads to a narrowing of the precipitation region of a nearly equimolar blockpolyampholyte poly(methacrylic acid)-*block*-poly(1-methyl-4-vinylpyridinium chloride) (PMAA-*b*-P1M4VPCl) [49]. A diagram of the states of PMAA-*b*-P1M4VPCl at various pH and ionic strength is shown in Fig. 6. The dissolution of PMAA-*b*-P1M4VPCl in the presence of KCl may be caused by the rupture of intraionic "bridges" between the oppositely charged blocks near the IEP. The appearance of the precipitation region at high

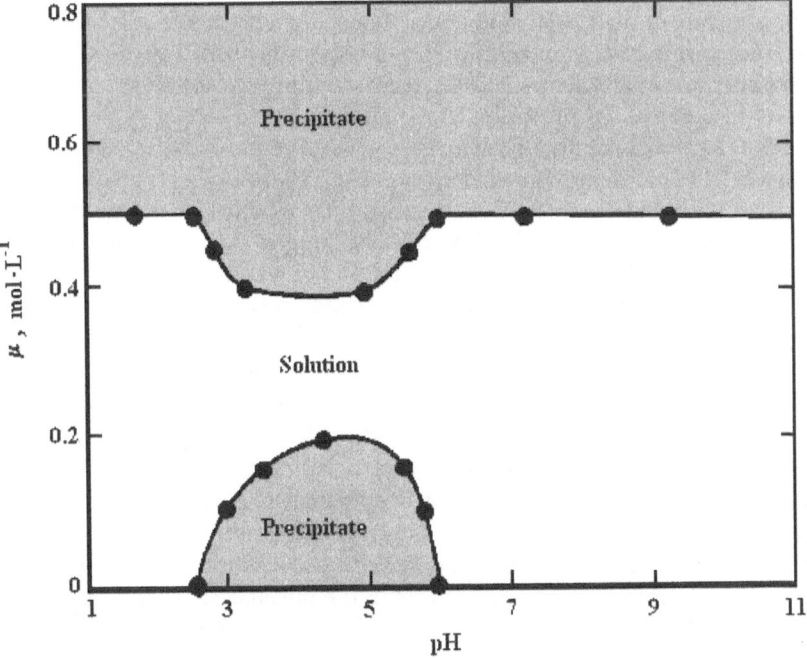

Fig. 6. Phase diagram of PMAA-*b*-P1M4VPCl (45:55 mol%) in aqueous salt solutions. C_{BPA}= 0.045 g·dL^{-1}

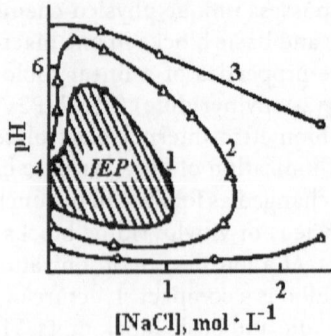

Fig. 7. Phase diagram of CCM-water-salt system as a function of the polymer concentration. C_p=0.05 (1), 0.1 (2) and 0.5 wt% (3)

ionic strength can be explained by the "salting out" effect. Figure 7 presents the phase diagram of chitosan carboxymethyl esters (CCM)/water/salt systems as a function of the polymer concentration, pH and ionic strength of the solution [50,51]. In the centre of the diagram, one can see an isoelectric region where the system is separated into two phases, one of which is solvent and the other a swollen precipitate of IPC. It can be seen from Fig. 7 that increasing the ionic strength up to 0.2–0.3 mol·L^{-1} causes the broadening of the boundaries of the heterophase nature of the system and hence impairment of CCM solubility. On a further increase in NaCl concentration, the region of phase separation becomes narrower and at μ >2.0 mol·L^{-1} the CCM solution with 0.1 wt% concentration can be prepared at any pH value. The solubility of triblock copolymer DMAEM–MMA–MAA (1:1:1) at the IEP, which is 6.6, is much lower than that one pH unit away [26]. By increasing the salt concentration, the solubility increases. At 0.9 M KCl the polymer is completely (at least 10 wt%) soluble even at the IEP. The increase in salt concentration leads to the screening of the attractions of acid and base blocks at the IEP and results in unfolding of macromolecules.

Triblock and random polyampholytes based on DMAEM–MMA–MAA were examined for their phase separation behaviour [52]. Triblock polyampholytes have a much broader phase separation region than the random ones. The specific structure of PMAA-*b*-P1M4VPCl with the excess of cationic or anionic blocks at the IEP is close to the structure of non-stoichiometric IPC. It is suggested that its nucleus consists of intraionic IPC surrounded by cationic blocks protecting it from precipitation [53]. ABC triblock copolymers of polystyrene-*block*-poly(2-(or 4)vinylpyridine)-*block*-poly(methacrylic acid) were synthesized by living anionic polymerization [53a]. Interpolymer complexation of the polyvinylpyridine and poly(methacrylic acid) blocks in the micellar solution was studied in relation to pH in solution by potentiometric, conductimetric and turbidimetric titration and in bulk by FTIR spectroscopy.

Light scattering data of random terpolymer and triblock copolymers DMAEM–MMA–MAA which have the same molecular weight 4000 and compo-

Table 5. Light Scattering Data for DMAEM–MMA–MAA (1:1:1) at pH = 5

Quasielastic light scattering			Static light scattering (Zimm plot)			
copolymer	d_h (Å)	Aggregation no.	M_w	A_2	d_g (Å)	Aggregation no.
random	40		5500			
triblock	110	21	125 000	+20.4	188	23

sition 1:1:1 are presented in Table 5 [26]. The molecular weight of the random terpolymer, as determined by static light scattering, is 5500±500, in reasonable agreement with the expected value of 4000, while the molecular weight of the triblock polymer, as determined by the same method, was found to be 125000±5000, which implies that aggregation occurs. The dependence of the hydrodynamic diameter of triblock polymer on pH shows that around the IEP big aggregates of size 100 nm are formed that are close to colloidal particles.

For self-concentrated aqueous solutions of polyampholytes it is possible to assume the existence of interparticle aggregates to account for the formation of high-ordered structures if polyampholytes are water-soluble at the IEP [54]. Raman spectra of equimolar copolymers of 1,2,5-trimethyl-4-vinylethynylpiperidol–4-methacrylic acid (TMVEP–MAA) at solid and liquid states are presented in Fig. 8. The spectrum of the polyampholyte in an amorphous state is very poor. The resolution of characteristic bands is not clear. Contrary, the resolution of characteristic bands is better for aqueous solution and the best at the IEP. Very sharp and narrow lines in the Raman spectrum, and especially the appearance of several lines in the low frequency region (60, 78, 87, 140 cm^{-1}), are evidence for the existence in aqueous solution of a high-ordered structure like crystals. How can such a structure be organized? When the amorphous copolymer is dissolved in water it takes the conformation of a Gaussian coil. Approaching the IEP compels the acidic and basic groups to interact and to form a globular structure stabilized by ionic contacts, hydrogen bonds and hydrophobic interactions. As a result very compact globular particles similar to the *secondary structure of proteins* are formed. These particles are protected from precipitation by zwitterions replaced on the surface of particles. At moderate concentration such globular particles tend to aggregate to minimize the surface energy (*tertiary structure*). At high concentration of polymer the aggregated macromolecules which are close to colloidal particles can be organized into the ordered crystalline structure. At the same time a small misbalance from the IEP to the acidic or basic region (pH=0.63±0.13) disturbs the ordered structure. The formation of a single crystal for colloidal silica particles as a result of "counterion-mediated attraction" was shown by Ise [55]. Figure 9 shows the SAXS curves for BSA at various pH including IEP [56]. When the solution is acidic or alkaline a peak is seen. However at the IEP (pH=5.06), the peak disappears, e.g. BSA forms an ordered structure in dilute solution.

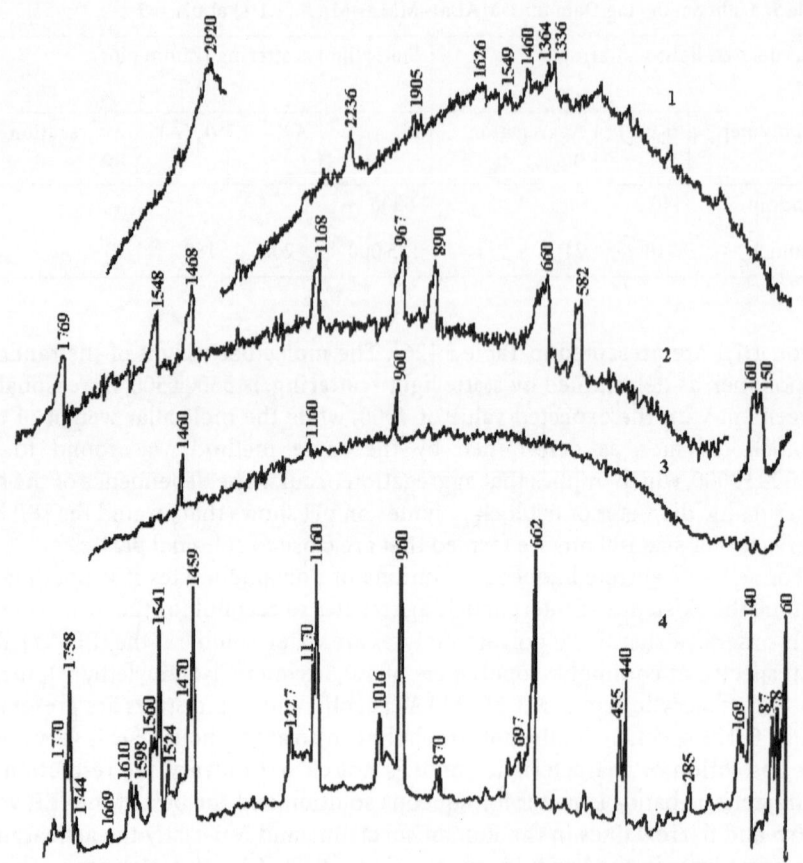

Fig. 8. Raman spectra of MAA–TMVEP in the solid state (*curve 1*) and aqueous solution (*curves 2–4*) at pH=7.5 (*curve 2*), 9.0 (*curve 3*) and 8.25 (IEP) (*curve 4*). C_P=5 g·dL^{-1}

It is also interesting to follow by the behaviour of hydrophobic polyampholytes. The influence of hydrophobic interactions on the solution properties of copolymers of α-olefins and mono-*N*-(3-dimethylaminopropyl)amide of maleic acid has been shown by Tanchuk et al. [32,33]. Especially unusual results were obtained from the dependence of the reduced viscosity on the temperature (Fig. 10). The reduced viscosity of PA-12 (see Table 1) increases with the increase in temperature and at T>340 K the solution transforms into a gel. The cooling of the system leads to the transformation of gel into the liquid state (sol-gel transition). Nature is successful with a large number of thermoreversible sol-gel transitions. For instance, agarose and gelatin undergo a gel-sol transition dependent on temperature. It is interesting to note that the model of amoeboid migration of cells on the surface is also associated with a cyclic sol-gel transition of the actin network [57] (Fig. 11).

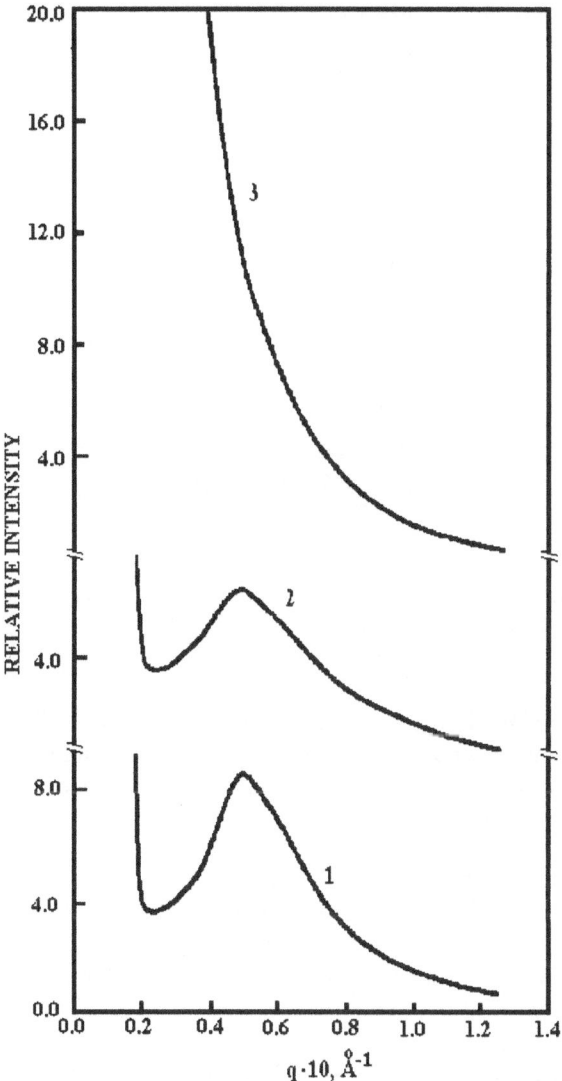

Fig. 9. SAXS curves of BSA at pH=10.94 (*curve 1*), 3.44 (*curve 2*) and 5.06 (*curve 3*). C_{BSA}= 0.075 g·dL^{-1} [55]

An appropriate balance of hydrophilicity and hydrophobicity in the molecular structure of polyampholytes is believed to be a key component in demonstrating the solubilization of organic probes and gelation properties [58]. The existence of hydrophobic clusters consisting of hydrophobic vinylbutyl ether (or styrene) fragments has been shown for alternating copolymers of vinylbutyl ether (or styrene) and *N,N*-dimethylaminopropylmonoamide of maleic acid

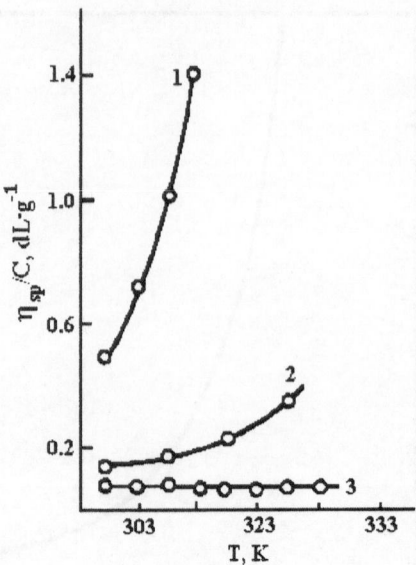

Fig. 10. Change in the reduced viscosity of hydrophobic polyampholyte in relation to the temperature. PA-12 (*curve 1*), PA-10 (*curve 2*), PA-8 (*curve 3*)

[59,60]. The maximal fluorescence intensity of 6-*p*-toluidino-2-naphthalenesulfonic acid was observed at the IEP of hydrophobic polyampholyte [59]. It can be suggested that at the IEP the long alkyl chains replaced inside of compact particles can form hydrophobic domain structures which enhance the solubilization of the organic probes. The macromolecules of hydrophobic polyampholyte PA-12 in aqueous solution near the IEP can also exist in the "helicogen" state by forming a hydrophobic core and a hydrophilic edge similar to natural polymers. At T >340 the formation of interpolyelectrolyte zwitterionic and hydrophobic contacts can take place which in turn cause the gelation of PA-12 (Fig. 12).

Table 6. Spin-Lattice Relaxation Time (T_1) and the Correlation Time (τ_c) Data of ^{14}N for 5 w/v% Aqueous Solutions at 25 °C

Polyam-pholyte	Without acids or salts		[HCl]/[Monomer]$_{av}$=1		KCl 1 M	
	T_1 (ms)	τ_c (ns)	T_1 (ms)	τ_c (ns)	T_1 (ms)	τ_c (ns)
PSN-25	5.5	0.63	4.3	0.80	4.5	0.77
PSN-50	6.0	0.57	5.5	0.63	4.6	0.75
PSN-75	7.3	0.47	5.5	0.62	4.6	0.75
PSN-100	8.0	0.43	7.2	0.48	6.5	0.53

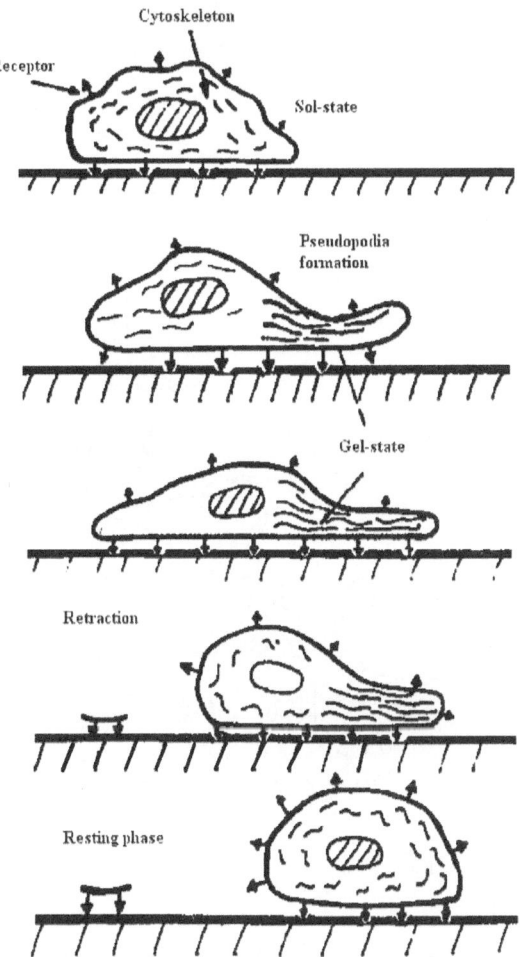

Fig. 11. Model of amoeboid migration of cells on the surfaces associated with cyclic sol-gel transition of the actin network [56]

In polyampholytes the interaction between charged macromolecules and surrounding low-molecular-weight electrolytes is important because the nature of these interactions plays an essential role in understanding the structure of polyampholytes on a molecular level. Authors [61] have reported the ^{23}Na, ^{35}Cl, and ^{39}K NMR relaxation rates and chemical shifts of amphoteric gel MAA-*co*-DMAMA under a variety of water contents and 1 M NaCl and 1 M KCl solution contents.

The dynamic mobilities of the cationic and anionic side chain of amphoteric copolymers of poly(sodium-2-methacryloyloxyethanesulfonate-*co*-2-metha-cryloyloxyethyltrimethylammonium iodide) (NaMES–METMAI) were estimat-

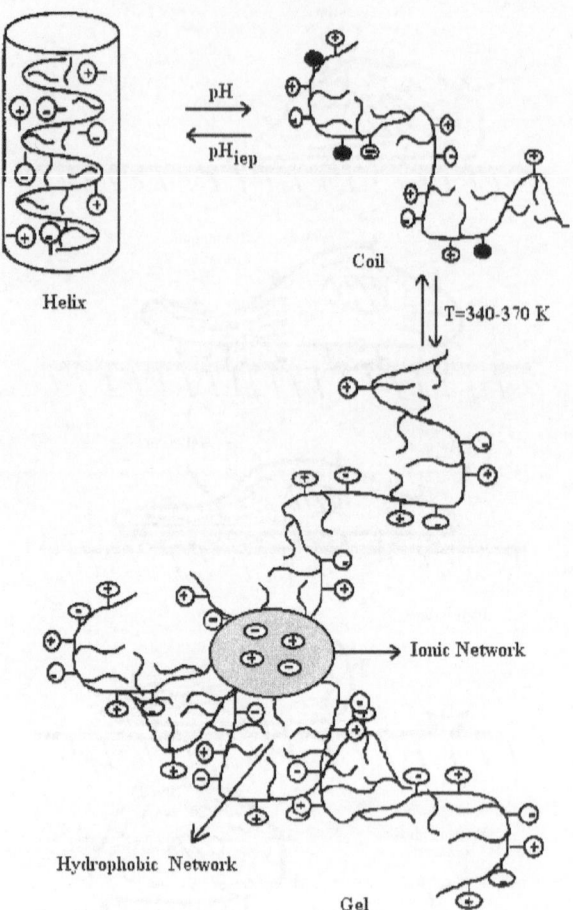

Fig. 12. Helix-coil and sol-gel transitions for hydrophobic polyampholyte PA-12 in relation to pH and temperature

ed with the quadrupolar ^{14}N NMR and ^{23}Na NMR relaxation techniques [62]. From Table 6 some important properties of the NaMES–METMAI copolymer series having the monomer feed ratios 3:1, 1:1, and 1:3 (named PSN-25, PSN-50, and PSN-75, where the numbers following PSN are the mole percents of MET-MAI in monomer mixtures) in aqueous solution can be ascertained. Firstly, T_1 times of ^{14}N decrease (τ_c becomes longer) as the component METMAI decreases. This means that the sulfonate component acts as a restriction environment to the trimethylammonium group by the attraction between them, and so slows down the motion of the trimethylammonium group. Secondly, τ_c of ^{14}N becomes longer when HCl or KCl is added to the aqueous solution of the PSN series. This phenomenon is due to the aggregation of the copolymers caused by the charge masking effect of an acid or a salt. The motion of sodium ions studied by Poisson–Boltzmann electrostatic theories and quadrupolar ^{23}Na NMR shows that PSN-0 and PSN-25 behave as typical polyelectrolytes.

6
Colloidal Properties of Polyampholytes

The main peculiarities of phase separation of polyampholytes – random copolymers of 2M5VP-AA that are insoluble at the IEP – were considered by authors [63–65] from the colloid chemical point of view. Tables 7 and 8 summarize some physico-chemical characteristics of polyampholytes used. The surface tension σ, viscosity [η] and electroconductivity χ of PA-3 solution and dispersion are minimal at the IEP (Fig. 13). In the semilogarithmic coordinates the dependence of surface tension σ on lgC has breaking points. They correspond to the critical concentration of polyampholytes C_{crit} that reflects the saturation of the adsorption layer on the liquid-gas boundary. Near the IEP the value of the critical concentration of polyampholytes corresponding to the limited value of σ is low, the surface activity G and the fraction of precipitated particles Ω are also maximal (Table 9). The dependence of C_{crit} on pH can be used to account for the deterioration of solvent quality. At $C=C_{crit}$ the macromolecular coils tend to float from the volume to the liquid-gas surface; at $C < C_{crit}$ the macromolecules are in the

Table 7. Composition, Intrinsic Viscosity, Weight Average Molecular Weight and pH_{IEP} of 2M5VP–AA

Abbreviation	[AA] (mol%)	[η] (dL·g^{-1})	$M_w \cdot 10^{-3}$	pH_{IEP}
PA-1	21.3	1.30	200	6.0
PA-2	27.1	0.73	111	5.9
PA-3	43.0	0.90	-	5.4
PA-4	53.5	0.38	93	5.0
PA-5	66.0	0.28	-	4.35

Table 8. Some Narrow Characteristics of PA-3 Fractions

Abbreviation	[AA] (mol%)	[η] dL·g^{-1}	$M_w \cdot 10^{-3}$	pH_{IEP}
PA-31	44.3	1.25	254.4	5.35
PA-32	42.8	0.89	134.3	5.42
PA-33	43.1	0.66	116.9	5.40
PA-34	44.2	0.38	55.1	5,35

Table 9. Influence of pH on C_{crit}, Surface Activity G and the Fraction of Precipitated Polyampholytes Ω^a

pH	5.1	5.2	5.4 (IEP)	5.5	5.75
C_{crit} (g·dL^{-1})	0.09	0.065	0.02	0.065	0.09
$G·10^3$ (J·m·kG^{-1})	45	180	1110	205	52
W	0.7	1.0	2.8	2.1	0,8

[a] $G=-(\partial\sigma/\partial C)_{C\to0}$; $\Omega=C_{PA}^d/C_{PA}^s$ where C_{PA}^d is the concentration of polyampholyte in the disperse phase; C_{PA}^s is the concentration of polyampholyte in solution.

compact and molecular dispersed state, at $C > C_{crit}$ the aggregation of compact particles takes place and the phase separation process occurs. It has been established that the average radius of particle r_λ, turbidity τ and the concentration of disperse phase are also maximal at the IEP of polyampholytes (Fig. 14). The dependence of the hydrodynamic diameter of triblock copolymer DMAEM–MMA–MAA (1:1:1) on pH appears in Fig. 15 [26]. Around the IEP big aggregates of size 100 nm form. At intermediate pH, smaller micellar aggregates of size 11 nm form. At extreme pH the strong electrostatic repulsions destroy the micelles and the polymer molecules occur in solution as separate chains.

The molecular weight and compositional heterogeneity of 2M5VP–AA copolymers having various molecular weights and composition have been studied by spectroturbidimetric titration [66]. The fractionation of copolymers in a mixture of methanol/toluene and methanol/diethyl ether leads to the precipitation

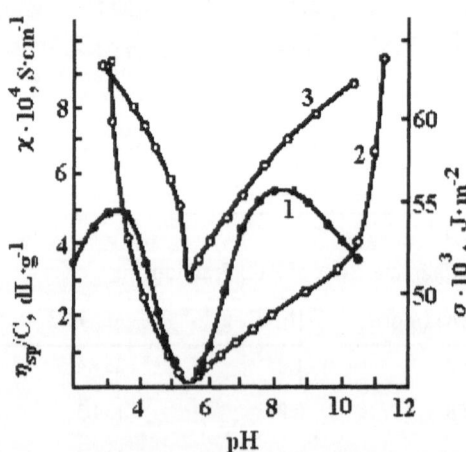

Fig. 13. Dependence of the reduced viscosity (*curve 1*), specific electroconductivity (*curve 2*) and surface tension (*curve 3*) of the solution and dispersion of PA-3 on pH. C_p=0.2 g·dL^{-1}

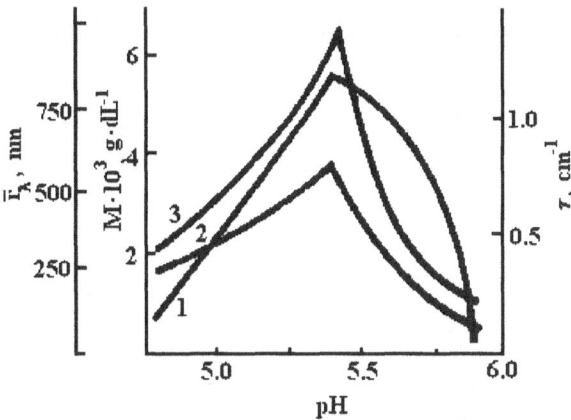

Fig. 14. Dependence of particles size (*curve 1*), concentration of precipitated polymer (*curve 2*), turbidity (*curve 3*) of PA-3 on pH of the solution. $C_p = 0.1$ g·dL^{-1}

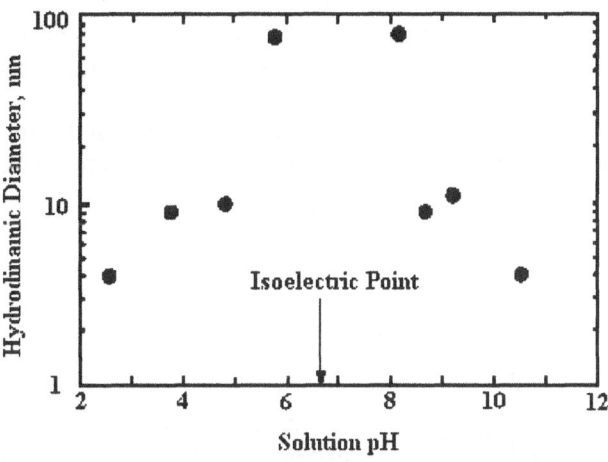

Fig. 15. Hydrodynamic diameter of triblock copolymer DMAEM–MMA–MAA (1:1:1) at different pHs with no added salt (Reproduced with permission from [26])

of polyampholytes by molecular weight and by composition, respectively. To establish the contribution of hydrophobic interactions to the stabilization of disperse particles near the IEP, the influence of simple electrolytes, dioxane, DMF and methanol has been studied. It has been shown that for stabilization of disperse particles of polyampholytes containing less than 25 mol% of AA groups, the hydrophobic interactions between methylvinylpyridine moieties are mostly responsible whereas for the other polyampholytes the driving force that stabilizes the disperse particles is the electrostatic attraction between acidic and basic monomers.

7

Association of Polyampholytes with Low-Molecular-Weight Ions

The complex formation problems of amphoteric copolymers with metal ions, detergents, dyes and organic probes is also of great interest. The interaction of metal ions M with polymeric ligands L proceeds as follow [67]:

$$M+L \Leftrightarrow ML; \; ML+M \Leftrightarrow ML_2; \; ML_{n-1}+M \Leftrightarrow ML_n$$

The stability constant of polymer–metal complexes β is equal to:

$$\beta_n = k_1 \, k_2 \, k_3 \, ... \, k_n = [ML_n]/[M] \, [L]^n$$

Function n, which characterizes the average number of ligands bound to metal ions, is determined from the equation:

$$n = \frac{[L]_t \, [LH] - [L]}{[M]_t}$$

where $[L]_t$ and $[M]_t$ are the total concentrations of ligand and metal ions; and $[L]$ and $[LH]$ are the concentrations of free (uncomplexed) and protonated ligands, respectively.

The concentrations of $[L]$ and $[LH]$ can be determined from the potentiometric titration curves of polymeric ligands in the absence and presence of metal ions. Figure 16 shows the complex formation curves for the system 2,5-dime-

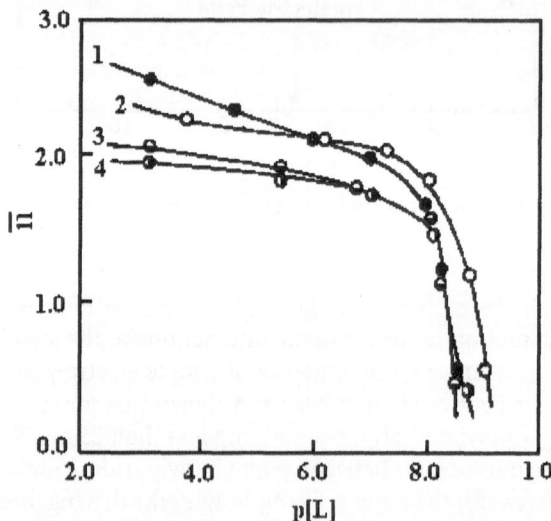

Fig. 16. Complex formation curves for the system DMVEP–AA/metal at μ=0.1 N $NaNO_3$. Cu^{2+} (*1*), Fe^{2+} (*2*), Co^{2+} (*3*), Ni^{2+} (*4*)

thyl-4-vinylethynylpiperidinol–4-acrylic acid (DMVEP–AA) and metal ions [68,69]. The number of monomeric units complexed with one metal ion is estimated as two from the plateau of the plot n vs. p[L]=–lg[L]. The stability of complexes is changed to $Cu^{2+} > Fe^{2+} > Ni^{2+} > Co^{2+}$.

Joanny and co-workers [70] presented a theory of the complexation of polyanions by divalent cations. The formation of both mono- and dicomplexes is possible. Monocomplexation locally inverts the charge of the polyelectrolyte and transforms it into a polyampholyte. The formation of a dicomplex creates a "bridge" between charged monomers.

Depending on the conditions polyampholytes are able to form several types of complexes and to involve either only acidic (or basic) or acidic and basic groups simultaneously. The chelate complex of Cu^{2+} with the maleic acid-*alt*-vinylamine includes one carboxylic and one amine group [71]. Multidentate polyampholytes, poly(*N*-propyleneglycine) (PPG), poly(1-isopropylcarboxylethyleneimine) (PIPCEI), poly(ethylenealanine) (PEA) are able to form five-membered chelate cycles [72]. The logarithm of complex formation constants lgβ of polyampholytes with respect to bivalent ions changes as follows: $Zn^{2+} > Ni^{2+} > Cd^{2+} > Co^{2+} > Pb^{2+} > Fe^{2+} > Ca^{2+} > Mg^{2+}$ and coincides well with the stability constants of EDTA (Table 10). 2M5VPy–AA can form several types of complexes with copper ions [73]. Analysis of the spectra indicates the existence of several types of complexes that are distinguished by the parameters $g_\perp$, $g_\parallel$ and $A_\parallel$. One, with $g_\perp=2.058$; $g_\parallel=2.334$ and $A_\parallel=136\cdot10^{-4}\,cm^{-1}$ corresponds to an A-type which is stable in the acidic region. The A-type complex belongs to cupric ions surrounded by two carboxylic groups with lgβ=7.1, but the line intensity of the A-type complex decreases when the pH is increased. In the neutral region pH=6–7.5, the appearance of the B-type complex with parameters $g_\perp=2.035$; $g_\parallel=2.298$ and $A_\parallel=157\cdot10^{-4}\,cm^{-1}$ occurs. The coordination number of the B-type complex in this region is equal to 4 with lgβ=21 that means two extra amino groups can be introduced into the coordination sphere of Cu(II) but the ESR lines of Cu(II) at pH=11.0 disappear. The behaviour of the system is reversible, e.g. the ESR lines appear again when the pH falls. The disappearance of an ESR signal can be ascribed to the formation of binuclear complexes stabilized by OH-bridges. The existence of "ESR undetectable copper" is also well known for copper-containing proteins ("blue" proteins).

Table 10. Stability Constants of Polyampholyte–Metal Complexes

Poly ampholyte	Mg^{2+}	Ca^{2+}	Fe^{2+}	Pb^{2+}	Co^{2+}	Cd^{2+}	Ni^{2+}	Zn^{2+}
PPG	7.2	8.9	17.0	17.4	18.8	18.6	19.0	20.8
PIPCEI	–	–	13.8	13.3	18.8	18.5	19.2	20.8
PEA	7.6	8.9	17.4	16.3	17.6	–	–	–

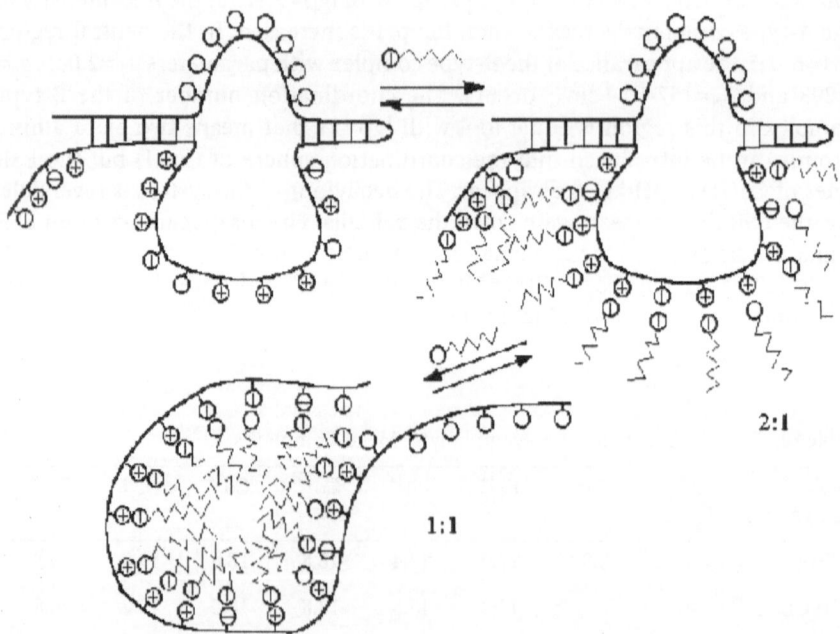

Scheme a

The stoichiometry of the vinyl ether of monoethanolamine–methacrylic acid/copper(II) (VEMEA–MAA/Cu^{2+}) has been established [74]. A maximum of the dependence of the optical density on the molar ratio of the initial components (Job method) indicates the formation of one type of complex in the pH interval 6–9. The following structure for polyampholyte-copper has been suggested:

Similar to the interaction of individual polyacids and polybases with surfactants, cationic detergents form cooperative complexes with acidic groups of polyampholytes, and anionic ones with their basic groups [75–78]. The addition of detergents leads to considerable variation in pH, electroconductivity, turbid-

Scheme 1. Successive formation of poycomplexes between PMAA*b*-P1M4VPCl and DDSNa

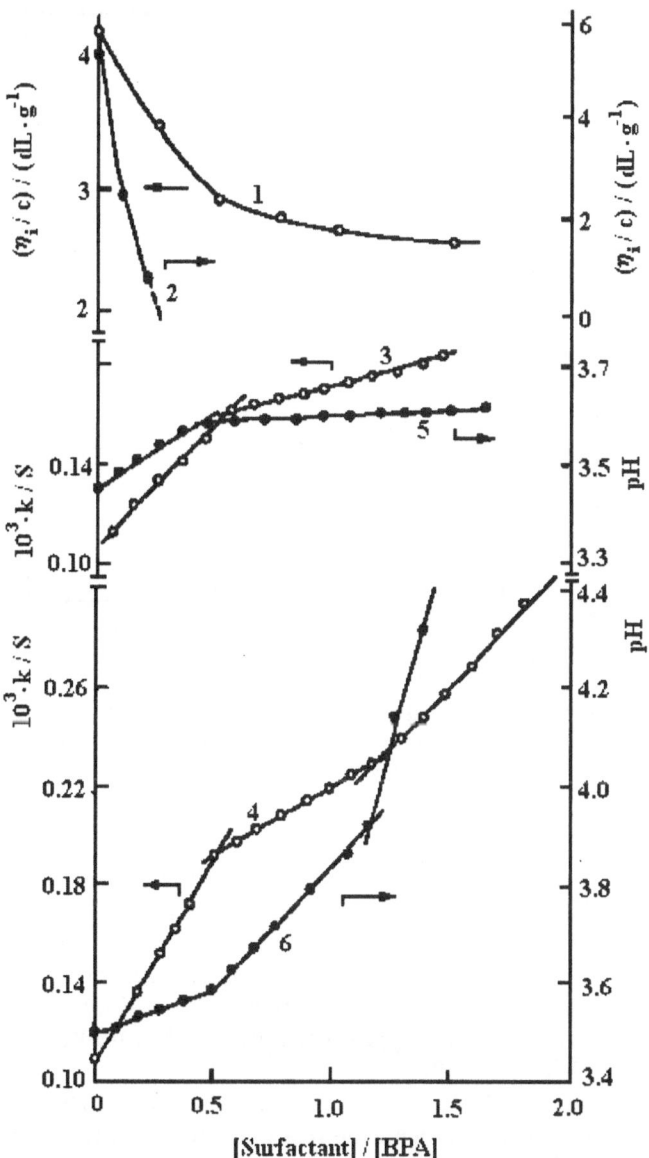

Fig. 17. Variation of reduced viscosity η_i/C (C_P=0.05 g·dL^{-1}) (*curves 1,2*), electroconductivity (*curves 3,4*) and pH (*curves 5,6*) on the composition of blockpolyampholyte/surfactant expressed as a mole ratio of repeating units of the active part of the blockpolyampholyte. [BPA]/[DDSNa] (*curves 1,3,5*), [BPA]/[CTMACl] (*curves 2,4,6*)

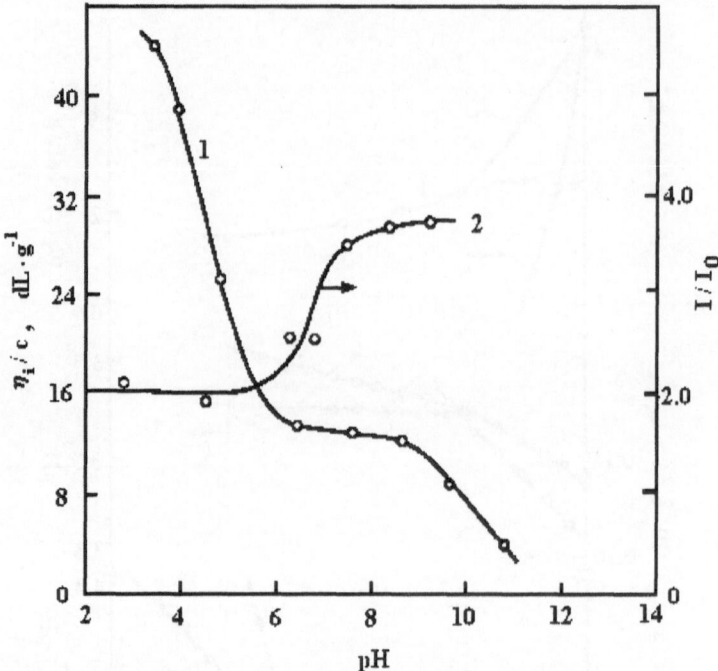

Fig. 18. Dependence of the reduced viscosity η_i/C ($C_P=0.05$ g·dL^{-1}) (*curve 1*) of block-polyampholyte and relative fluorescence intensity I/I_0 of ANS (*curve 2*) on the pH of the solution. [ANS] = $5 \cdot 10^{-5}$ mol·L^{-1}, $\lambda=320$ nm

ity and viscosity of the system. This is connected with the formation of compact particles stabilized by hydrophobic contacts of the long alkyl chain of detergents. These particles are preserved in solution with the help of noncomplexing components of copolymers. The potentiometric and conductimetric titration curves of nearly equimolar blockpolyampholyte (PMAA-*b*-P1M4VPCl) by DD-SNa shows the formation of two types of complexes [PMAA-*b*-P1M4VPCl]/[DD-SNa]=2:1 and 1:1 (Fig. 17) [79].

The formation of [PMAA-*b*-P1M4VPCl]/[DDSNa]=2:1 can be accounted for by the interaction of DDSNa with the free uncomplexed parts of the cationic block. The formation of the stoichiometric complex is due to the destruction of cooperative ionic contacts between acidic and basic blocks and replacement of them by new cooperative ionic contacts. The successive formation of polycomplexes between PMAA-*b*-P1M4VPCl and DDSNa can be represented by Scheme 1. For the system consisting of PMAA-*b*-P1M4VPCl and CTMACl the formation of [PMAA-*b*-P1M4VPCl]/[CTMACl]=2:1 only is observed. In this case the cationic groups of the surfactant cannot compete with the cationic groups of the blockpolyampholyte and interact only with the carboxylic groups placed on the "loops". Figure 18 shows the variation of relative fluorescence intensity I/I_0 of 8-anilino-1-naphthalenesulfonic acid (ANS) and of the reduced viscosity of PMAA-b-P1M4VPCl vs. pH of the solution. A sharp decrease of the reduced viscosity over a small pH interval can be accounted for by the increase of intramacromolecular polyelectrolyte complex formation. The considerable increase of I/I_0 of ANS over pH 6–8 is probably the result of incorporation of dye molecules into the hydrophobic region of intramacromolecular complex.

8
Interpolymer Complexes of Polyampholytes

By now we have rather rich theoretical and experimental materials accumulated for interpolyelectrolyte complexes stabilized by cooperative ionic contacts and interpolymer complexes stabilized by hydrogen bonds. Little attention, however, has been paid to polymer–polymer reactions involving synthetic polyampholytes. The complexes of 2M5VPy-AA with PAA have been studied by various physico-chemical methods [80]. It has been established that the formation of a common cooperative system consisting of ionic and hydrogen bonds between the copolymer and PAA is realized. Authors [81] have considered the interpolyelectrolyte reaction between sulfo derivatives of chitosan – polyampholyte, containing amino and sulfo groups, and 2,5-ionene bromide. The turbidimetric titration curves of 2,5-ionene bromide by chitosan sulfate at pH=2.5 and 11.5 show that the composition of PEC is quite different in acidic and alkaline regions (Fig. 19). In acidic medium only half of the sulfate groups participate in the formation of the nonstoichiometric complex because the rest exist in the form of intramolecular PEC. In the alkaline region the composition of PEC is stoichiometric because all the sulfate groups are involved in the intermolecular PEC formation reaction. The competition between intra- and intermolecular ionic contacts is clearly seen from the comparison of ^{13}C NMR spectra of MDAA–MA and a mixture of MDAA–MA and PDMDAAC at pH=3.9 corresponding to pH$_{iep}$ of the alternative polyampholyte [82] (Fig. 20) Peaks at 180.5 and 180.8 ppm correspond to carboxylate anions of MDAA–MA involved in intramolecular complexes while the shifting of the same peaks up to 181.3 and 181.8 ppm can said to account for the formation of intermolecular complexes between carboxylate anions of MDAA–MA and quaternary nitrogen atoms of PDMDAAC. At the same time at the IEP of DMDAA–MA (pH$_{IEP}$=4.1), the strength of intramolecular contacts between quaternary ammonium and carboxylate groups is higher than that between the intermolecular ones because the position of carboxylate anions of DMDAA–MA at 182.4 ppm does not change during the addition of PDM-DAAC.

As distinct from the statistical and alternative ones, blockpolyampholytes are able to bind both polyacids and polybases as well as non-ionic hydrophilic polymers more cooperatively [83,84]. Figure 21 shows the potentiometric, conductimetric and viscometric curves of titration of PMAA-b-P1M4VPCl by PAA, PVBTMACl, PVPD and PEG. The composition of PEC was found from the bend of the curves. All PECs are water-soluble because they are protected from precip-

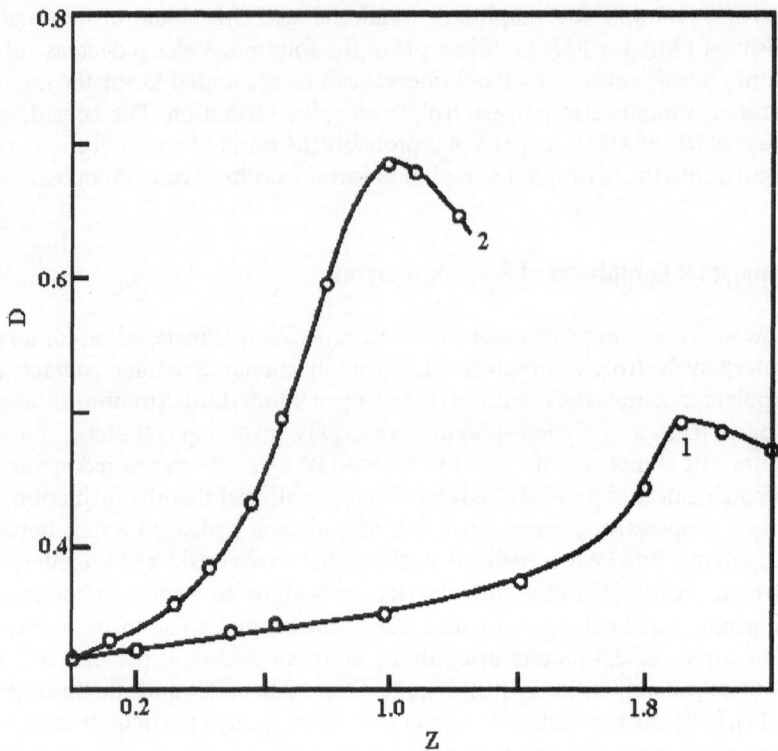

Fig. 19. Turbidimetric titration curves of 2,5-ionene bromide solution by chitosan sulfate (CS) at pH= 2.5 (*curve 1*) and 11.5(*curve 2*). [2,5-ionene]=5 · 10^{-4} mol·L^{-1}. Z=[CS]/[2,5-ionene]

itation with the help of free unloaded parts of PMAA-*b*-P1M4VPCl. The phase diagram of the system PMAA-*b*-P1M4VPCl/PVBTMACl in relation to PEC concentration and pH of the solution is seen from Fig. 22. At C_{PEC} >0.075 g·dL^{-1} PEC has a wide region of insolubility at the interval of pH=2–12. The opaque solution exists at pH >5.0 and C_{PEC} <0.075 g·dL^{-1}. The microgel structure is formed preferentially in the acidic region. Thus we have for this system three phases: solution, gel and precipitate which can be transformed into each other by changing either the PEC concentration or the pH of the solution. This behaviour of PEC can be explained by Scheme 2. At relatively high C_{PEC}, PEC particles aggregate and precipitate. At relatively low C_{PEC}, they are dispersed and form an opaque solution. The phase transition precipitate/microgel and opaque solution/microgel at low pH can be explained by the swelling of PEC particles as a result of suppression by the ionized carboxylic groups of the acidic block which are responsible for PEC formation. The gel formation process is accompanied by the release of a certain amount of PMAA "loops" which form interchain networks with the help of H-bonds.

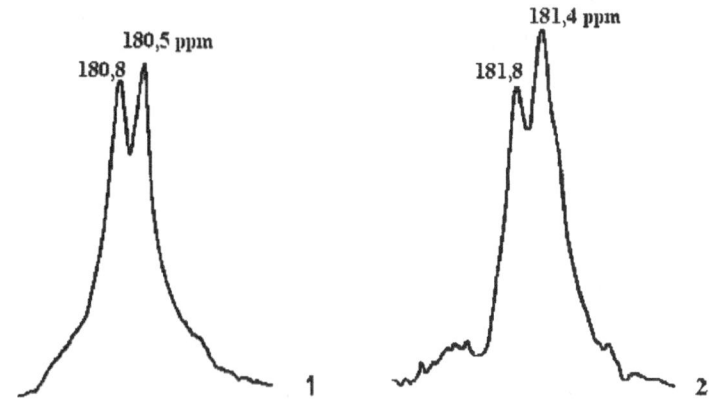

Fig. 20. ^{13}C NMR spectra of carboxylic groups of MDAA–MA (*1*) and a mixture of MDAA–MA and PDMDAAC (*2*) at pH=3.9 that corresponds to the pH$_{IEP}$ of MDAA–MA

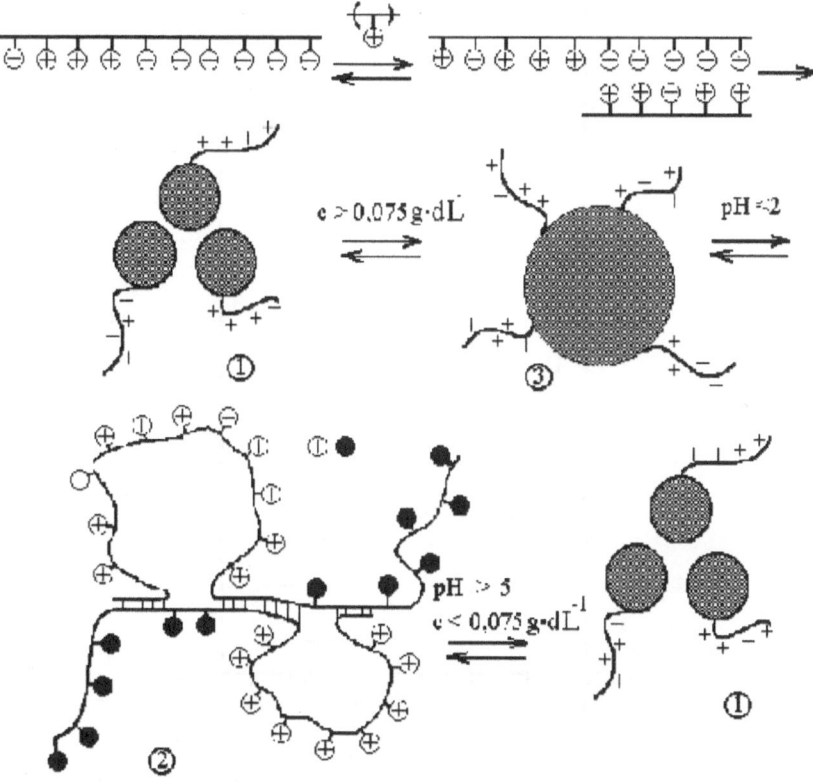

Scheme 2. Behaviour of polyelectrolyte complexes

The turbidimetric titration curves of 1-methyl-4-vinylethynylpiperidinol–4-methacrylic acid (MVEP–MAA) by HSA in aqueous solution at various pHs are shown in Fig. 23 [85]. The maximal values of turbidity correspond to the composition of polyampholyte–protein associates expressed as $n=m_{HSA}/m_{PA}$ (where m_{HSA} and m_{PA} are the mass fraction of HSA and the polyampholyte, respectively). As seen from Fig. 24, the most favorable region of interaction of HSA

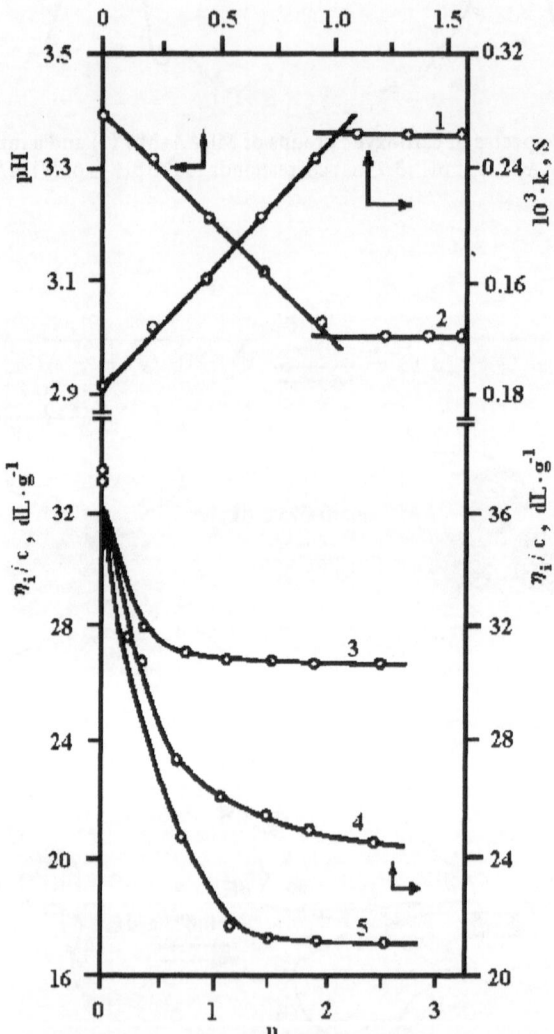

Fig. 21. pH-metric (*curve 1*) conductimetric, (*curve 2*), and viscometric (*curves 3–5*) titration of BPA by PAA (*curves 1,2*), PEG (*curve 3*), PVBTMACl (*curve 4*) and PVPD (*curve 5*). n=[BPA]/[homopolymer]

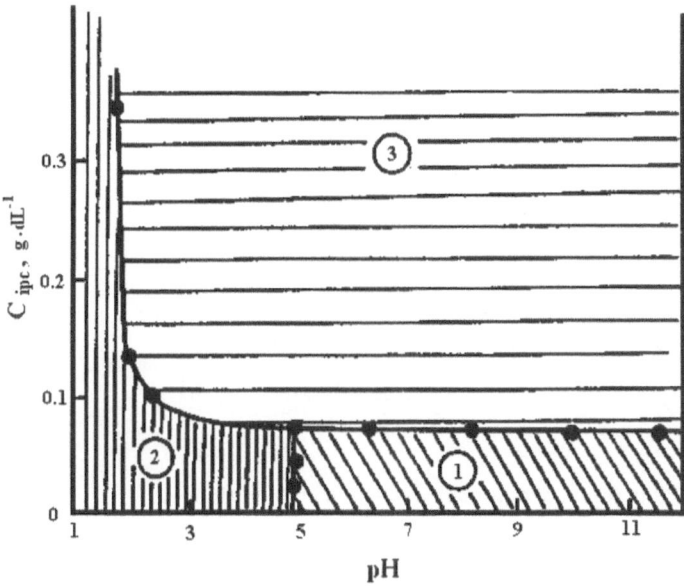

Fig. 22. Phase transition of PEC consisting of BPA and PVBTMACl in water. Opaque solution (*1*), microgel (*2*), precipitate (*3*)

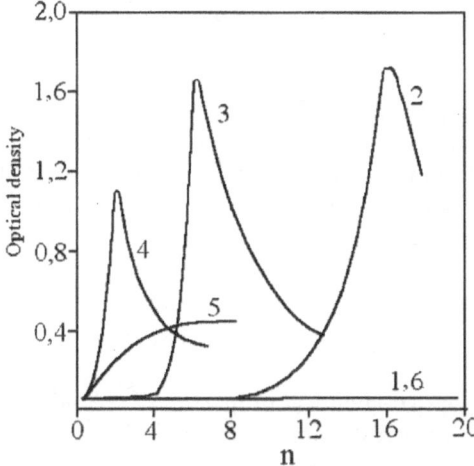

Fig. 23. Turbidimetric titration curves of MVEP–AA by HSA in water at pH=5.1 (*1*), 6.2 (*2*), 7.1 (*3*), 8.5 (*4*), 9.1 (*5*), 10.0 (*6*). $n=m_{HSA}/m_{PA}$

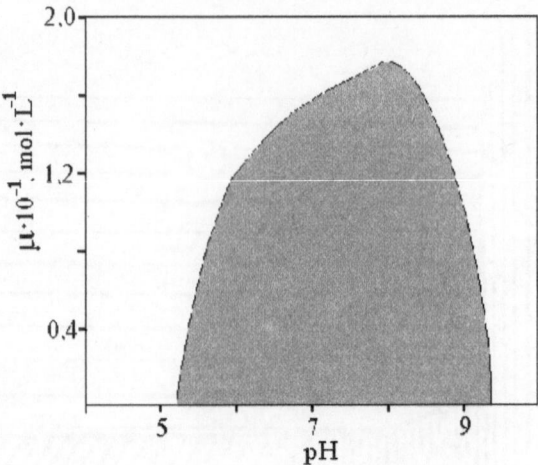

Fig. 24. Stability region of polyampholyte–protein associates

(pH_{IEP}=5.4) and MVEP–MAA (pH_{IEP}=9.4), as also noted by other authors [86], arranges between their isoelectric points. Increasing the ionic strength of the solution makes the association interval of pH narrower.

The complexation behaviour of proteins with dilute solutions of PAA and a random polyampholyte DMAEM–AA–MMA was studied by turbidimetric titration [87]. Polyampholyte–polyampholyte interaction (self-aggregation) and polyampholyte–protein complexation was studied as a function of pH and polymer dosage. Large increases in turbidity were observed for polyampholyte–protein mixtures compared with polyampholyte alone. However, protein analysis of the supernatant and precipitate revealed that only about 10% of the protein precipitates with the random polyampholyte while 90% of the protein remains in the equilibrium liquid. An experiment with PAA and oppositely charged protein shows the opposite trend with 90% precipitation of protein. Hence, great care needs to be taken in the interpretation of turbidimetric titration data. It has been reported that polyampholyte-protein associates behave with lowered immunogenic activity [88].

9
Adsorption of Polyampholytes on Disperse Particles

Regulation of the physico-chemical properties of colloid dispersions by polyelectrolytes is of great interest from the theoretical and practical points of view [89]. A consideration of the peculiarities of absorption of polyelectrolytes on disperse particles can be of help to clarify some aspects of kinetics and mechanism of flocculation, elemental acts of flocculation, as well as to provide the se-

Table 11. Survey of Polyampholytes Employed

Polyampholyte	$Mw \cdot 10^{-3}$	Mole ratio acid/base	pH_{IEP}	Solubility at the IEP*
AA/2M5VP	170	1:1	5.1	S
	–	3:1	4.2	I
AA/VI	20	1:1.3	5.4	S
AA/DMVEP	200	1:1	4.1	S
	100	2:1	4.5	S
MAA/TMVEP	250	2.5:1	5.0	l
PMAA/P1M4VPC	200	1:1.2	5.5	I

*) S: Soluble; I: Insoluble.

lective flocculation of polydisperse particles. The adsorption of HSA, ribonuclease [90,91] and immunoglobulin (IG) [92] on the polystyrene latexes (PSL) was investigated. It has been shown that the maximal adsorption is observed at the IEP of proteins. The influence of synthetic polyampholytes on the stability of cells *Candida lambica* has been investigated [93]. It was established that the decrease of optical density of the system is the result of aggregation of cells due to the simultaneous adsorption of charged macromolecules on several cells. An

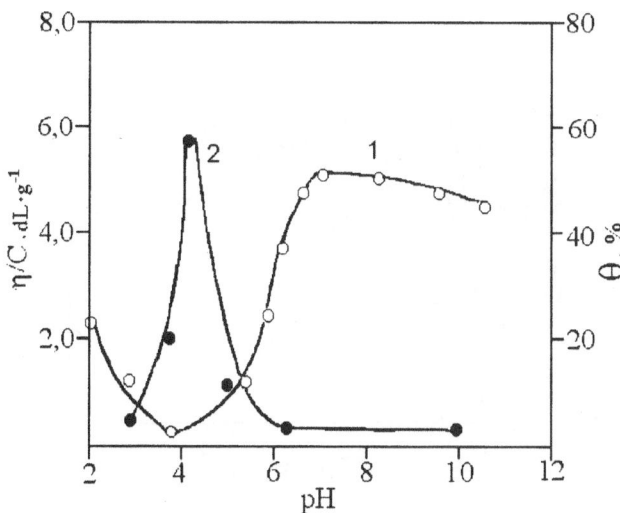

Fig. 25. Dependence of the reduced viscosity of AA/DMVEP (*curve 1*) and the binding degree (θ) of PSL by AA/DMVEP (*curve 2*) on pH of the solution. [AA/DMVEP]=4.5 · 10^{-3} mol·L^{-1}

equation which describes the kinetics of the coagulation process has been found. However, concerning the interaction of synthetic polyampholytes with disperse particles, there is no available information in the literature except for experiments performed by Blaackmeer [94] and Ouali et al. [95]. Joanny [96] has studied theoretically the adsorption of a single random polyampholyte chain on a charged planar surface mimicking the latex particle. His concluding remarks are in good agreement with our experimental results performed recently on the interaction of synthetic polyampholytes with colloid dispersion [97]. Table 11 represents some characteristics of polyampholytes applied for flocculation of disperse particles.

Figure 25 shows the influence of polyampholyte conformation on the binding degree of PSL at various pH. The maximal binding degree takes place at the IEP of AA–DMVEP and coincides well with the behaviour of proteins mentioned elsewhere [90–92].

The dependence of the optical density (A) of PSL on the concentration of various types of polyampholytes is presented in Fig. 26. The flocculating effectiveness of polyampholytes is arranged as follows: PMAA-b-P1M4VPCl >AA–VI >MAA–TMVEP >AA–DMVEP and is in good agreement with the sequence of their electrokinetic potential (Fig. 27). The kinetics of flocculation of PSL allow the evaluation of the magnitude of the retardation factor W which characterizes

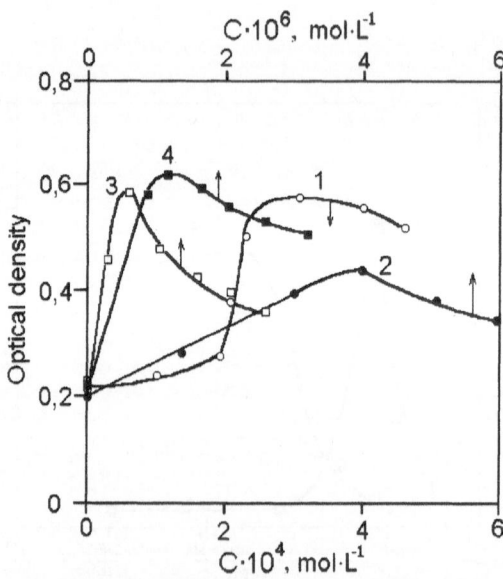

Fig. 26. Dependence of the optical density (D) of PSL on the concentration of added polyampholytes: AA/DMVEP (*curve 1*), MAA/TMVEP (*curve 2*), PMAA-b-PMVP (*curve 3*), AA/VI (*curve 4*)

the decrease in the rate of process in comparison to the fast coagulation and is related to the potential energy of interaction U by Eq. 9.1 [98]:

$$W = 2a \int_{2a}^{\infty} \exp(U/kT) R^{-2} dR \qquad (9.1)$$

where a is the radius of particles; k is the Boltzmann constant; T is the absolute temperature; and R is the distance between the particles centres. The factor W can be calculated from the optical density A by Eq. 9.2 [98]:

$$(dA/dt)_{t \to 0} = \frac{B k_0 C_L^2}{2,303 W \rho^2} = \frac{k' C_L^2}{2,303 W} \qquad (9.2)$$

where ρ is the density of latex particles; C_L^2 is the concentration of latex particles in $g \cdot mL^{-1}$; B is the optical constant; k_0 is the rate constant of fast flocculation and $k' = B k_0 / \rho^2$.

Since for the same latex particles the values of B, k_0 and ρ are constant, the magnitude of k' can be determined from Eq. 9.2 at W=1, e.g. in the region of fast flocculation, while k' is constant for each flocculant and does not depend on the concentration of the system. As seen from Fig. 28 the retardation factor W changes over a wide range of polyampholyte concentration and the theoretical

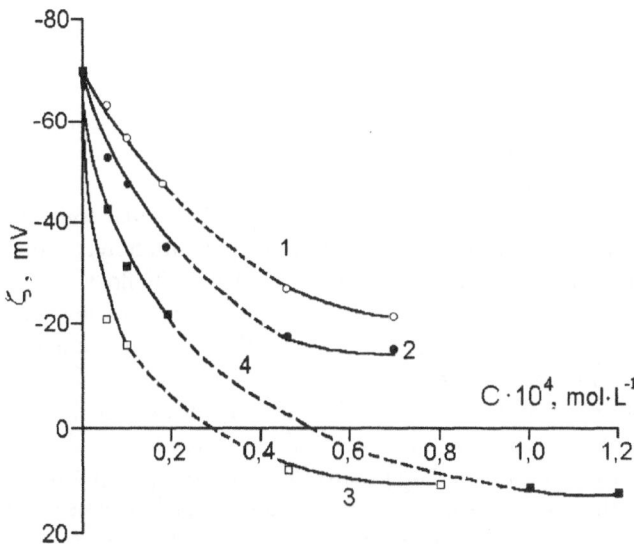

Fig. 27. Variation of the electrokinetic potential of PSL particles on the concentration of polyampholytes: AA/DMVEP (*curve 1*), MAA/TMVEP (*curve 2*), PMAA-*block*-PMVP (*curve 3*), AA/VI (*curve 4*). The dotted lines correspond to lack of stability of PSL

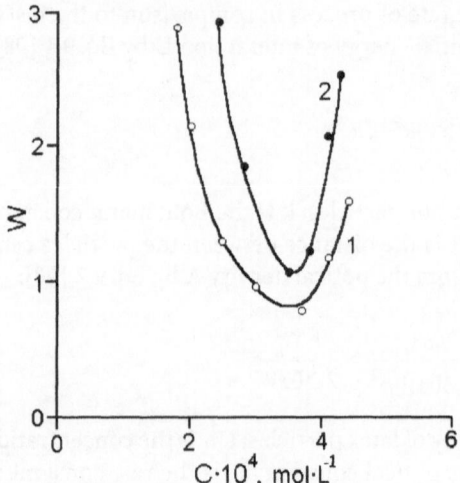

Fig. 28. Dependence of the stability factor of PSL (W) on the concentration of AA/DMVEP.
1 – experimental results, *2* – theoretical calculations

curve does not fit with the experimental one. These results show that another mechanism can be realized that leads to the destabilization of the disperse system in spite of the high energy of the double electric layer which stabilizes the system by electrostatic mechanism. It has been supposed that the flocculation of PSL can be explained by formation of "bridges" between latex and polyampholyte particles [99]. At the same time it is also necessary to take into account the possibility of heterocoagulation as a result of unproportional adsorption of polyampholytes on the surface of PSL leading to the appearance of nonproportional charges and potential on the latex particles.

A comparison of surface area of PSL and macromolecular coils, accepting that the latter is spherical at the IEP, shows that the polyampholyte molecules in dependence of their nature can theoretically involve in the flocs about 28–60 latex particles (Table 12). However, in the case of polyampholytes, the amount of latex particles involved in the flocs is much higher than that theoretically calculated. These data cannot be explained either by DLVO theory [100] or "bridge" or heterocoagulation mechanisms [101]. The mechanism of adsorption of polyampholytes on latex particles can successfully be explained in the framework of theoretical results performed by Joanny [96]. According to Joanny, the polyampholyte solution can adsorb even if its overall charge has the same sign as that of the latex particles. Two cases where the interaction between functional groups of polyampholytes is short range (strongly screened) and long range (unscreened) are considered. For short-range interactions, Joanny, using the replica trick and a Hartree approximation, showed that the adsorption of the chain is

Table 12. Aggregation of PSL by Polyampholytes at Their pH_{IEP} [a]

Floc culant	Flocculation effectivity (%)	Number of flocs		Degree of aggrega- tion	Flocculating concentration $10^{-4}mol \cdot L^{-1}$
		theoretically calculated	experimen- tally found		
AA/DM- VEP (1:1)	94	60	154	5.6	6–10
MAA/TM VEP (1:1)	65	36	1452	2.2	0.2–0,4
AA– 2M5VP (1:1)	92	28	1044	1.8	0.3–0,5
AA– 2M5VP (1:3)	82	32	664	3.4	2–4,7

[a] The numerical concentration of latex particles is equal to $4.63 \cdot 10^{17}$ particles/m^3

possible even if it has the same net charge as the surface and if the interaction potential is larger than a critical value, i.e.:

$$V_c = \log\frac{f}{g} \tag{9.3}$$

where V_C is the adsorption threshold; and f and g are the fraction of monomers having, respectively, the same charge and the opposite charge as the surface. If the overall charge of the adsorbing chain is opposite to that of the surface, there is no adsorption threshold ($V_C < 0$). When the total charge is small, which is realized at $f/g \approx 1$, the value of V_C is small. When $f=g$ there is no adsorption threshold, the chain always adsorbs even though the average potential acting on the monomers vanishes. The latter is interpreted as follows: when the monomers oppositely charged to the surface monomers are in contact with the surface, the monomers having the same charge as the surface do not feel the repulsive potential which is assumed here to have an infinitely short range. In other words, for short-range interactions, the redistribution of the charges of the polyampholyte on the surface of charged disperse particles takes place and an entropy penalty must be overcome. It is important to note that, even if the chain is on average repelled by the surface ($f > g$), it can adsorb if the interaction potential V is large enough. Taking into consideration the concluding remarks of Joanny, one can propose the mechanism of interaction of polyampholyte solutions with PSL and suggest the next structure of latex–polyampholyte associates.

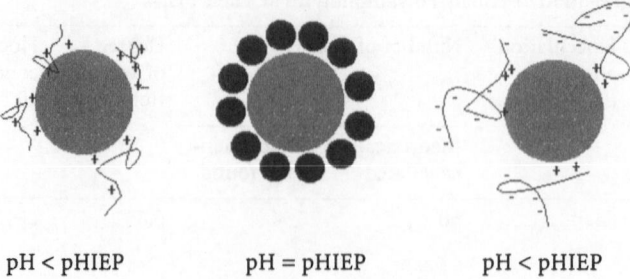

      pH < pHIEP   pH = pHIEP   pH < pHIEP

Scheme b

The decrease of flocculation effectiveness of polyampholytes on both sides of pH_{IEP} with respect to negatively charged latex particles can be explained by the adsorption of positively charged parts of polyampholytes that protect the floc formation owing to the presence of positively or negatively charged loops dangling on the surface of particles. The continuous contact of polyampholyte molecules (concentration is equal to $0.1\ g \cdot L^{-1}$) with PSL (size of PSL is 120 nm) from 15 min to 24 h increases the value of specific adsorption (Γ) of polyampholytes from 1 to 10 $mg \cdot g^{-1}$. According to Gregory [101] the interaction of polymers with dispersions consists of several stages: a) distribution of macromolecules among the particles; b) adsorption of polymer segments on the surface of particles; c) redistribution of adsorbed chains, e.g. the transition of macromolecular conformation from the initial state to equilibrium; and d) the collision of complex particles with floc formation. Table 13 summarizes the kinetic parameters of adsorption of polyampholytes calculated according to Gregory [101]:

$$k_{1,2} = (2kT / 3\eta)(r_1 + r_2)^2 / r_1 r_2 \tag{9.4}$$

where $k_{1,2}$ is the adsorption constant rate; k is the Boltzmann constant; T is absolute temperature; η is the viscosity of the medium; and r_1 and r_2 are the radius of the latex particles and the polyampholyte coils, respectively. The time t_A that is necessary for adsorption of polyampholyte, f, can be find from Eq. 9.5 suggesting that the concentration of latex particles in the system N_1 is constant and the adsorption rate constant $k_{1,2}$ does not depend on the degree of covering of the surface by polyampholytes.

$$t_A = -\ln(1 - f) / k_{1,2}\ N_1 \tag{9.5}$$

Equation 9.5 represents a rough approximation because both the number of particles and the sum of surface should decrease in the course of adsorption leading to a decrease of $k_{1,2}$. Nevertheless, it allows the calculation of the minimal time which is necessary for adsorption of part of the polyampholytes. From Table 13 it follows that the time taken to adsorb 90% of the polyampholyte reaches 32 s; 50% and 10%

Table 13. Kinetic Parameters of Adsorption of DMVEP–AA by PSL Calculated according to Eqs. 9.4 and 9.5

r_1 PSL (m)	r_2 PA (m)	N_1 particles/m^3	$k_{1,2}$ m^3/s	f (%)	t_A (s)
$6 \cdot 10^{-8}$	$2.1 \cdot 10^{-7}$	$4.63 \cdot 10^{17}$	$1.56 \cdot 10^{-19}$	90	32
				50	9,6
				10	1,46

take 9.6 and 1.4 s, respectively. But the calculated values of t_A are much less than those experimentally observed. For instance to reach the specific adsorption of polyampholytes corresponding to approximately half of Γ_∞ on latex with r=60 nm (mg·g^{-1} over 24 h) it is necessary to spend 20 min. Even to reach 10% adsorption more than 3 min is needed, which is significantly higher than the theoretically predicted value of Γ. Thus the adsorption of polyampholytes by latex particles is not determined by the diffusion of macromolecules to the surface. It could be that the redistribution of adsorbed macromolecules in relation to time is the main factor leading to the displacement of small disperse particles by large ones. The diffusion process proceeds faster and is completed in several seconds dependent on the size of the particles while the redistribution process proceeds slower (from several minutes to

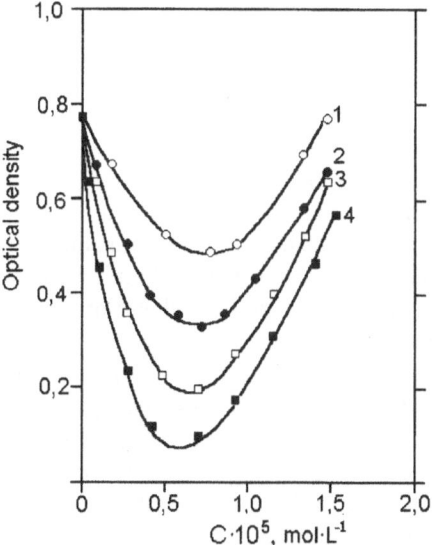

Fig. 29. Dependence of the optical density of barite suspension on the concentration of AA/2M5VP with pH$_{IEP}$=4.2 (*curves 1–3*) and AA/2M5VP with pH$_{IEP}$=5.1 (*curve 4*). The fixed time from the precipitation of dispersion is 10 (*curve 1*), 20 (*curve 2*) and 30 min (*curves 3,4*)

several hours) and leads to significant growth of specific adsorption. The marked peculiarities are to be taken into account during the flocculation of latex particles by polymers when the contacting time of macromolecules with dispersions is limited by minutes. The stability of mineral suspension – barium sulfate in the presence of AA/2M5VP – passes through the minimum (Fig. 29). In contrast to latex particles, barite suspension has a rough dispersity. The minimal concentration of polyampholytes causing the flocculation of $BaSO_4$ corresponds to $3–6 \cdot 10^{-5}$ mol·L^{-1}. The addition of polyampholytes at first leads to the aggregation of mineral particles and sedimentation. Therefore, the optical density of the system gradually decreases. A further increasing of the optical density corresponds to the formation of stable disperse particles. At the same time the optimal concentration of one-, two- and three-valent cations (Na^+, Ca^{2+}, Y^{3+}) leading to flocculation of mineral suspension is equal to $20 \cdot 10^{-4}$, $8 \cdot 10^{-4}$ and $1 \cdot 10^{-4}$ mol·L^{-1}, respectively. The optimal size of flocs can be formed only at a definite ratio of concentration of polymers and solid phase. Therefore the dependence of the sedimentation velocity of particles on the concentration of polymers is extremal.

10
"Forcing Out" Phenomenon at the Isoelectric Point of Polyampholytes

At the IEP electrostatic attraction forces between oppositely charged groups are so high that they are condensed into a compact globule squeezing out the solvent. "Force out" phenomenon taking place only near the IEP was at first noted by our research group and afterwards was confirmed by several scientific centres. The main idea of this phenomenon comes from the fact that any low- or high-molecular-weight substances associated with polyampholytes can be released at the IEP as a result of a competition between inter- and intramolecular interactions. In other words if the cooperativity of intrachain interaction of acidic and basic groups within a single macromolecule predominates over those for interchain interaction such phenomenon can be realized. A simple example of this effect is the detachment of amino acids at their zwitterionic or isoionic states from the surface of ion-exchange resins that are applied for separation and analysis of amino acid sequences of proteins. Amphoteric ion-exchangers also act in the same way during the desorption and elution of preliminary absorbed low- and high-molecular-weight ions from the polymeric matrix. However, the realization of "forcing out" phenomenon for a series of water-soluble polyampholytes has been demonstrated by our research group. We have found that these types of polyampholytes are able to bind metal ions, dye molecules and polyelectrolytes at definite pH values and release them quantitatively at the IEP [102–104]. Sorption and desorption processes of transition-metal ions by such polyampholytes can be represented as follows (Scheme c):

If the IEPs of polyampholytes are displaced in the alkaline region, transition metal ions are precipitated in the form of hydroxide and the polyampholyte is found in the solution and can be recovered and used several times. It is interesting to note that a similar effect was observed for a ternary poly(acrylic acid)–

Scheme c

copper(II)–poly(ethyleneimine) system (PAA–Cu^{2+}–PEI) [105]. At definite pH, the displacement of nitrogen–metal coordination bonds and carboxylate–metal ionic bonds by the cooperative polyacid–polybase contacts can probably lead to the precipitation of polyelectrolyte complexes and retaining of metal ions in the supernatant as shown below (Scheme d):

Earlier [106–108] the amphoteric character of the PEC particle dispersion with the IEP, characterized by zero mobility and a minimum in the reduced viscosity, was shown.

The dependence I/I_0 of 1-anilino-8-naphthalensulfonic acid (ANS) and acridine yellow (AY) bound to DMVEP–AA (pH$_{IEP}$=7.0) on the pH of the solution is shown in Fig. 30. The maximal value of I/I_0 for ANS at pH=4 is the evidence for the binding of negatively charged dye molecules with cationic parts of polyampholyte. A marked decrease of I/I_0 for ANS and sharp increase for AY near the IEP, approaching those for the ANS and AY solutions alone, can be explained by the release of dye molecules from the macromolecular coils of the polyampholyte. The existence of "forcing out" phenomenon was clearly shown for systems consisting of polyampholyte and anionic or cationic polyelectrolytes [102]. The composition of PEC determined from the breaks of the potentiometric, conductimetric and turbidimetric curves is equal to [polyampholyte]/[polyelectrolyte]=3:1.

Figure 31 shows the pH dependence of the reduced viscosity for the individual components and PEC. Phase separation of products composed of polyam-

Ternary PAA–Cu^{2+}–PEI complex   →   PEC precipitates

Scheme d

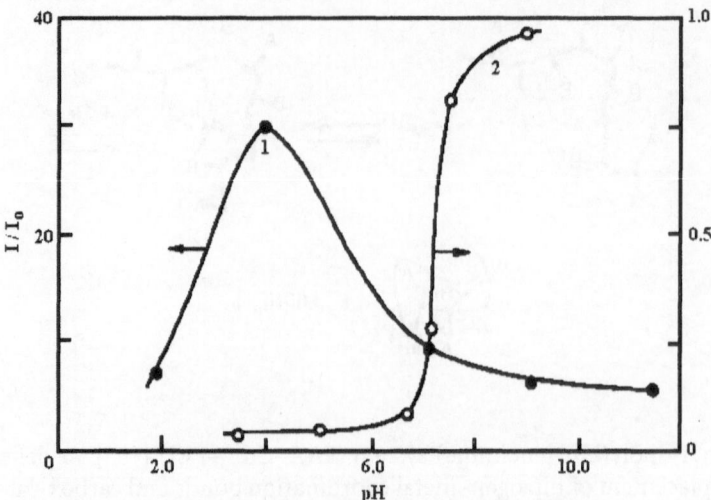

Fig. 30. Dependence of the fluorescence intensity (I/I_0) on pH for the systems polyampholyte–ANS (*curve 1*) and polyampholyte–AY (*curve 2*)

pholyte (DMVEP–AA) and polyacid (PAA) occurs at 3.0< pH <5.5; for PEC consisting of polyampholyte (DMVEP–AA) and cationic polyelectrolyte (PDM-DAAC), this process takes place at pH >8.2. Attention should be paid to the fact that near the IEP of DMVEP–AA ($pH_{IEP}=7.0$), the reduced viscosity of PEC is sharply increased. The only reasonable explanation of this fact is the destruction of PEC particles into individual components. Since the value of reduced viscosity of DMVEP–AA is much lower, near a pH=7, the main contribution to the viscosity should come from the viscosities of anionic or cationic polyelectrolytes.

To confirm a cooperative destruction of PEC particles near the IEP, independent sedimentation experiments were carried out. Figure 32 shows the sedimentation diagrams of DMVEP–AA, PAA and PEC at various pH. The sedimentograms of PEC show one peak with S=4.75 and 4.18 at pH=6.1 and 6.9, respectively. However at pH=7.1 two peaks with S=1.9 ("slow" peak) and 5.6 ("fast" peak) appear on sedimentograms. At the same time the sedimentation coefficients of separate components DMVEP–AA and PAA at pH=7.2 are equal to 3.53 and 1.06, respectively.

The appearance of slow and fast sedimentation peaks at a narrower interval of pH ($\Delta pH=0.2$) change is the result of PEC particle destruction into the individual macromolecules. The structure of PEC composed of polyampholyte and polyelectrolyte can be represented as double-strand sequences of pairs formed with the help of cooperative systems of ionic and hydrogen bonds. Probably near the IEP, some acidic and basic groups of polyampholytes displaced on loops begin to interact with each other cooperatively with the formation of intramolecular complexes. The mechanism of PEC destruction can be represented as follows:

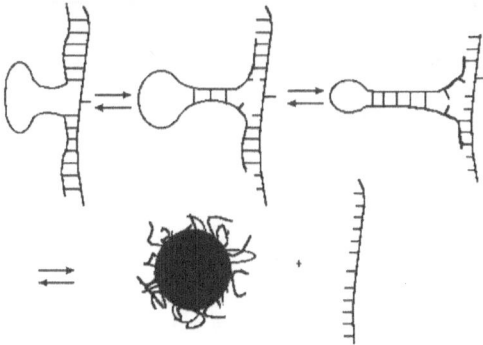

Scheme e

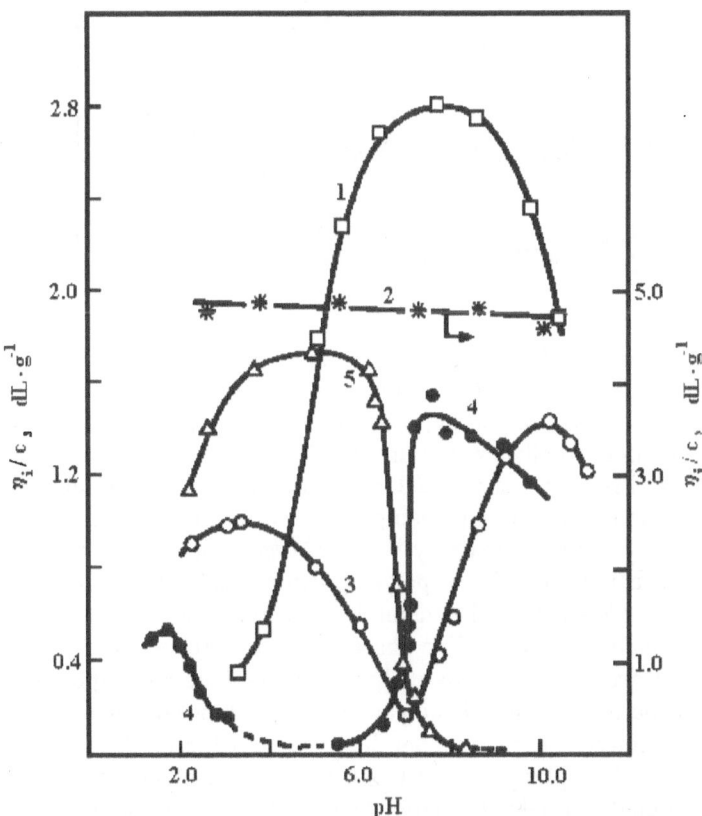

Fig. 31. Influence of pH on the reduced viscosity of PAA (*curve 1*), PDMDAAC (*curve 2*), DMVEP–AA (*curve 3*) and polyelectrolyte complexes composed of DMVEP–AA/PAA (*curve 4*), DMVEP–AA/PDMDAAC (*curve 5*) in water

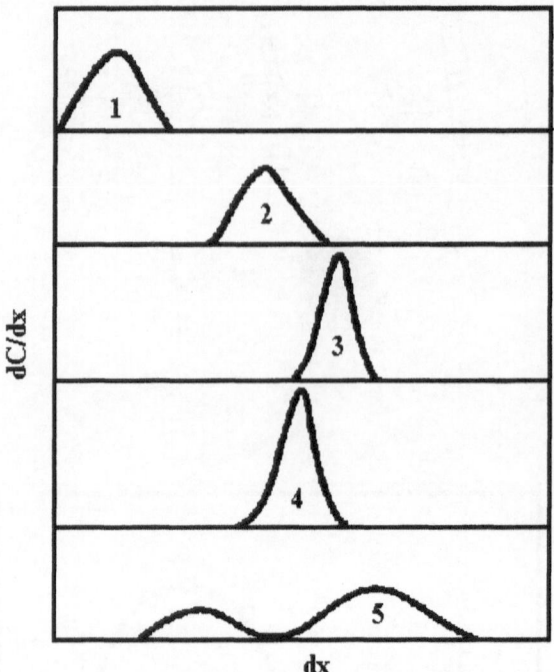

Fig. 32. Sedimentograms of DMVEP–AA (*curve 1*), PAA (*curve 2*) and polyelectrolyte complexes composed of [DMVEP–AA]/[PAA]=3:1 (*curves 3–5*) at various pH in aqueous solution

The dependence of PEC yield on pH shown in Fig. 33 can also serve as additional confirmation of validity of the proposed mechanism [109]. The separate titration curves of polyampholyte by sodium poly(styrenesulfonate) (SPSS) and poly(N,N-dimethyldiallylammonium chloride) (PDMDAAC) intersect at the IEP where the yield of PEC is also equal to zero. Recently, the realization of the "forcing out" effect in relation to polyampholyte–surfactant complexes was demonstrated [110]. In contrast to the system considered above in the case of polyampholyte–surfactant complexes, polyampholytes (2M5VP–AA, DMAEM–MAA and DEAEM–MAA) at the IEP themselves precipitate and surfactant molecules (CTMA) remain in solution. The phase formation process or "phase rejection" proceeds as avalanche-like. An analysis of precipitate and supernatant shows that the disperse phase is only polyampholyte and the soluble part is pure surfactant. It is interesting to note that the precipitated (after destruction of polycomplexes) polyampholyte 2M5VP–AA becomes insoluble in the alkaline region although the initial polyampholyte itself shows a good solubility in alkaline medium. In other words the "forcing out" effect probably causes the rearrangement of the initial conformation of polyampholyte. The insolubility of polyampholyte particles in the alkaline region can be accounted for by the disposition of preferentially vinylpyridine groups on the surface of aggregates which, after

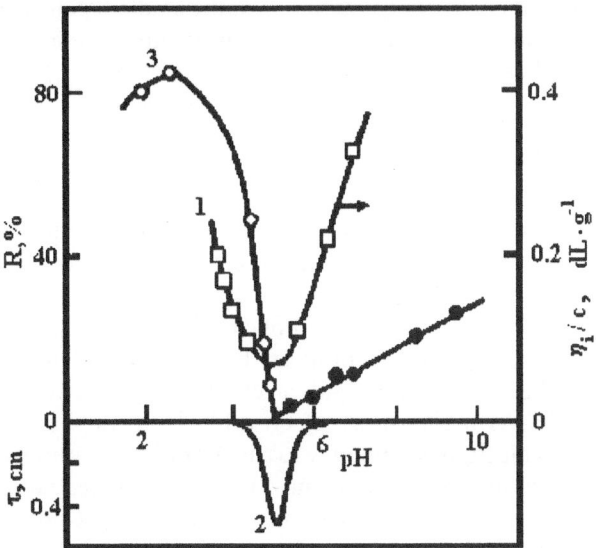

Fig. 33. Dependence of the reduced viscosity (*curve 1*), turbidity (*curve 2*) of TMVEP–MAA and yield of polyelectrolyte complexes (R, %) composed of TMVEP–MAA and PSS (or PD-MDDAC). *Open circles* correspond to titration of TMVEP–MAA by PSS. *Closed circles* correspond to titration of TMVEP–MAA by PDMDAAC

Table 14. Comparative Phase Formation of 2M5VP–AA Copolymers at the IEP

AA (mol%)	2M5VP (mol%)	pH_{IEP}	$M_w \cdot 10^{-3}$	C_P (g·dL^{-1})	Ξ_1^*	Ξ_2^{**}
44.3	55.7	5.35	254.4	0.1	4.0	62.9
42.8	57.2	5.42	134.3	0.1	2.4	54.9
43.1	56.9	5.40	116.9	0.1	0.9	53.1
44.2	55.8	5.35	55.1	0.1	0.7	52.4

[*] Degree of phase formation of polyampholytes before surfactant adsorption;
[**] Degree of phase formation of polyampholytes after release of surfactant molecules at the IEP

Table 15. Comparative Phase Formation of DMAEM–MAA Copolymers at the IEP

MAA (mol%)	DMAEM (mol%)	pH_{IEP}	C_P (g·dL^{-1})	Ξ_1	Ξ_2
20.5	79.5	5.5	0.1	-	52.2
42.3	57.7	5.3	0.1	-	62.3
62.3	37.7	4.3	0.1	10.0	75.1

the dissolution of the same polyampholyte in the acidic region, become soluble in the alkaline region as usual and take an initial conformation. The comparison of the degree of phase formation of polyampholytes before (Ξ_1) and after release of surfactant (Ξ_2) shows that the value of Ξ_2 is much higher than Ξ_1 independent of the molecular weight of copolymers (Tables 14 and 15).

One can suggest that during the interaction of polyampholytes with surfactant molecules the latter prearrange the macromolecular chain so that more suitable conformation of polyampholytes and more profitable arrangement of acidic and basic groups will be provided in order to promote and intensify the cooperative formation of both intra- and interchain contacts between oppositely charged groups. Thus, in the case of polyampholyte–surfactant systems, the mechanism of "molecular recognition" can be realized. The "forcing out" effect can successfully be used in matrix polyreactions [104]. Scheme 3 illustrates the synthesis of poly(N-vinylpyrrolidone) on a blockpolyampholyte matrix. The idea of applying a blockpolyampholyte as a matrix offers advantages in comparison with polyacids and polybases alone: 1) matrix polymerization can be carried out on separate blocks; 2) water-soluble products can be protected from the precipitation by free (non-loaded) acidic or basic blocks; 3) easy separation of "daughter" chains from the matrix can be realized as a result of competitive inter- and intramacromolecular complex formation reactions; 4) a narrow molecular weight product can be produced; and 5) the blockpolyampholyte can be re-used several times.

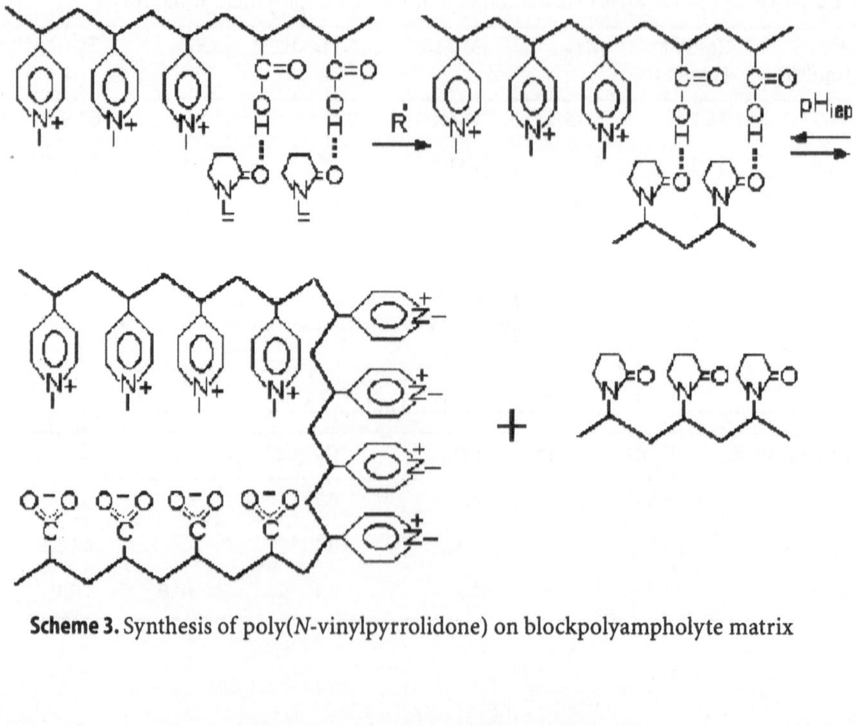

Scheme 3. Synthesis of poly(N-vinylpyrrolidone) on blockpolyampholyte matrix

11
Properties of Zwitterionic Copolymers

Polymeric betaines or zwitterions are polyampholytes whose oppositely charged groups remote one from another are displaced on one pendant substituent. There are several types of monomers with a betaine structure: carboxybetaines, sulfobetaines and phosphobetaines. Poly-N-ethyleneglycine (1), poly(N-3-sulfopropyl)-N-methacryloyloxyethyl-N,N-dimethylammonium betaine) (2) and poly[(2-methacryloyloxyethyl-2-(trimethylammonioethylphosphate)] (3) are typical examples of this kind of polyampholyte:

$$-CH_2-CH_2-NH^+- \quad ; \quad -CH_2-\underset{\underset{CO}{|}}{\overset{\overset{CH_3}{|}}{C}}- \quad ; \quad -CH_2-\underset{\underset{CO}{|}}{\overset{\overset{CH_3}{|}}{C}}-$$

Compound 1 branch: CH_2, COO^-

Compound 2 branch: CO, $(CH_2)_2$, O, $+N(CH_3)_2$, $(CH_3)_2$, SO_3^-

Compound 3 branch: CO, $(CH_2)_2$, PO_2, O, $(CH_2)_2$, $N^+(CH_3)_3$

$$\textbf{1} \qquad\qquad \textbf{2} \qquad\qquad \textbf{3}$$

The synthesis and solution properties of polymeric betaines were outlined in detail by Salamone and co-workers [111–121], in particular in the vinylimidazole and vinylpyridine series, as well as by other research groups [122–128]. Recently [129–131], the preparation and solution properties of polybetaine, containing the phosphatididylcholine group, poly[(2-methacryloyloxy)ethyl-2-(trimethylammonioethyl phosphate)] which has PO_4^- and $-N^+(CH_3)_3$ groups divided by two methylene groups have been reported. The linear relationship between $[\eta]$ and $1/\mu$ was observed for the polyampholyte at low μ near pH_{IEP}. When μ is >0.0025, $[\eta]$ increases due to the release of the attractive interactions between the oppositely charged units. By comparing the μ dependence of the electrostatic expansion factor of the polyampholyte with that of poly(sodium acrylate), it is

Table 16. Hydrodynamic and Molecular Characteristics of Poly(N-3-sulfopropyl)-N-methacryloyloxyethyl-N,N-dimethylammonium betaine) in Water-Salt Solutions

Samples	SB-8	SB-9	SB-10	SB-11	SB-12	SB-13
$M_w \cdot 10^5$	11.4	12.2	15.0	26.9	36.7	50.1
C_s (mol·L^{-1})			R_g/R_h (nm)			
0.06	44.6/34.7	47.0/35.8	50.6/39.8	70.2/52.6	81.6/63.0	97.1/72.0
0.08	44.3/36.2	46.7/37.9	52.0/41.8	71.0/65.7	87.5/70.8	100.5/82.0
0.1	45.6/37.0	47.4/38.8	53.6/43.1	80.4/67.4	93.3/80.1	113/95.1
0.3	47.3/39.0	49.7/42.0	56.0/47.0	81.5/70.0	99.8/84.5	128/102.7
1.0	47.0/40.5	49.0/42.5	56.5/47.8	89.5/72.0	101/91.3	135/105.0

suggested that there is a pronounced intramolecular attraction between the oppositely charged segments even when pH deviates from the IEP.

Polymeric betaines are usually insoluble in pure water and have gel characteristics but are soluble in salt-containing solutions. The loss of water solubility and gel-like structure that adopts polybetaines are probably due to the formation of intra- and interchain ion contacts which result in the appearance of cross-linked networks. The intrinsic viscosity [η], second virial coefficient A_2, exponent a in the MKH equation, the radius of hydration R_g and the hydrodynamic radius R_h increase with the increase in salt concentration C_s [132] (Table 16).

The exponents of the MKH are equal to a=0.5; 0.67; 0.70 and 0.70 for C_s=0.06; 0.3; 1.0 and 4.0 M NaCl aqueous solutions, respectively. The electrostatic expansion factors for polyampholyte effect, α_e, were estimated for the sulfobetaine polymers over a wide range of molecular weights and C_s. It is concluded that the chain expansion for a neutral polyampholyte is controlled by the non-ionic excluded volume effect and the electrostatic excluded volume effect (polyampholyte effect) at moderate added-salt concentration. The electrostatic expansion factor can reasonably be described by a α_e^3-type equation, although no such an equation for the polyampholyte has so far been proposed.

Solubility of polybetaines in aqueous solution depends on the nature of the anions and cations of the added salts. For salts having a common anion (Cl$^-$) and one-valent cations, an increase in solubility changes as follows: Li$^+$ >NH$_4^+$ >Na$^+$ >K$^+$, while for two-valent cations the solubility increases in the order: Ba^{2+} >Sr^{2+} >Ca^{2+} >Mg^{2+}. In the presence of salts with a common cation (K$^+$), but different anions, the solubility increases in the following order: ClO$_4^-$ >I$^-$ >Br$^-$ >Cl$^-$. Table 17 represents the minimal concentrations ($C_s \cdot 10^{-2}$ mol·L^{-1}) of various salts needed to dissolve poly[N,N'-diethyl(acrylamidopropyl)ammoniumpropanesulfonate] (PDMAAPS) [133]. Charge/radius ratio, Hoffmeister lyotropic sequences and Pearson theory can successfully be applied to explain the solubility behaviour of polybetaines.

Table 17. Influence of the Nature of Cations and Anions on the Solubility of PDMAAPS

	Cl	Br	I	ClO_4	NO_3	NO_2	SO_4	CH_3COO
Li	4.26							
NH_4	4.09							
Na	3.60			0.77		2.86	4.48	22.11
K	3.57	2.08	0.94	0.86	2.86			
Mg	2.81							
Ca	2.44							
Sr	2.27							
Ba	2.08							

The synthesis [134] and characterization [135,136] of aromatic and aliphatic poly(sulfonatopropylbetaines), together with the binding ability with respect to optical fluorescent and chemically reactive anionic organic probes [137], have been described. The strongly dipolar character of the zwitterionic structure of polybetaines affords a number of specific properties: a) "antipolyelectrolyte" behaviour as typified by increasing chain expansion with increasing ionic strength of the aqueous solution; b) strong binding capacity, with simultaneous dehydration, toward a variety of organic anionic probes in aqueous solution; and c) high solvation power of the glassy polymeric matrixes toward salts of widely different lattice energy leading to completely amorphous blends [138]. Hydration of acrylic (methacrylic) polyzwitterions bearing aliphatic quaternary ammonium-sulfonatopropylbetaine as a side group ($\geq N^+-(CH_2)_3-SO_3^-$) and poly(2-vinyl- and 4-vinylpyridinium sulfonatopropylbetaines) (P2VP-SB and P4VP-SB, respectively) have been studied over a broad range of water content (weight fraction <0.5) [139].

Comprehensive analysis of the first-order transitions and heat capacity of sorbed water, at glass transition temperature (T_g) has allowed the identification of the main common features of the system: a) at 296 K water diffusion is Fickian and the isotherm analysis shows that water sorption occurs according to a multilayer sorption process [140]. It allows discriminating site-bound water (1.5–2.0 water molecules per monomeric unit) and indirectly bound water without great differences of binding energy between these successive solvation layers. Moreover, clustering of water molecules is never observed, not even for the highest water concentrations; b) differential scanning calorimetric measurements [141] allow the quantification of two different types of bound water: non-freezable bound water (type I), 7.7±0.9 molecules per monomeric unit, showing mobility fairly similar to that of bulk water and a strong plasticizing effect on the system; freezable bound water (type II), 6.7±0.9 molecules per monomeric unit, characterized by a complex multipeak melting endotherm occurring in a rather broad temperature range (242–272 K); and c) the poly(pyridinium zwitterions) display slightly decreased hydrophilicity with respect to their aliphatic quaternary ammonium homologues, e.g. a significantly lower number of non-freezable and freezable bound water molecules per monomeric unit (type I), 5.0 and (type II) 5.5±0.5 showing a two-peak melting endotherm at 267±1 K and 271±1 K.

According to molecular mechanics calculation and dielectric studies [142], triethylammonium-, -pyridinium- and 2,6-dimethylpyridiniumsulfonatopropyl betaines are characterized in dilute aqueous solution by a nearly fully extended conformation with fairly similar intercharge distances (4.8–5.0 Å) leading to dipole moment values of about 23.0–24.2 D. It may also be stressed that the aqueous solutions of aromatic and aliphatic poly(zwitterions) exhibit significantly different phase behaviours: P2VP-SB has an upper critical solution temperature (UCST) at 286 K, poly[N,N-dimethyl-N-[2-(methacryloyloxy)ethyl]-n-(3-sulfonatopropyl)ammonium betaine] shows both UCST at 306 K and an "apparent inverted" lower critical solution temperature (LCST) at 289 K [124,143]. [see also conflicting data of Huglin MB, Radwan MA (1991) Polymer International 26: 97] and P4VP-SB is insoluble over the whole temperature range between 273 and 373 K.

Recently, Laschewsky et al. [144–149] have reported on the synthesis and characterization of zwitterionic polysoaps which combine the advantages of the behaviour of polyzwitterions and micellar polymers. The variation of the polymer geometry produces "head-type" (1), "mid-tail type" (2) and "tail-end type" (3) zwitterionic polysoaps.

Scheme f

The viscosity of the cationic polysoap exhibits polyelectrolyte character whereas the viscosity of the zwitterionic polysoaps behaves as an uncharged polymer. The unusually low viscosities of zwitterionic polysoaps are attributed to the intramolecular aggregation of the hydrophobic side chains, keeping the hydrodynamic radius small. The surface activity and solubilization capacity of amphoteric polysoaps have been studied. The sequence to solubilize hydrophobic dyes is "mid-tail" > "head" > "tail-end" geometry.

The dielectric properties of mixtures of the inorganic salt NaI and zwitterionic polymethacrylates were measured in the frequency range from 10^2 to 10^7 Hz and at temperatures between 110 and 400 K for different salt concentrations (0, 100 and 200 mol%) [150]. One relaxation process is observed whose relaxation rate depends strongly on the length of the aliphatic spacer between the polymethacrylate main chain and the zwitterionic group. Exhibiting an Arrhenius-like temperature dependence with activation energy $E_A = 47$ kJ/mol, this relaxation process is assigned to fluctuation of the quaternary ammonium group in the side chains. At higher temperatures, the dielectric properties and the conductivity are primarily dominated by the mobile inorganic ions: conductivity strongly depends on the salt concentration, showing a pronounced electrode polarization effect. The frequency and salt concentration dependencies of the conductivity can be quantitatively described as hopping of charge carriers being subject to spatially randomly varying energy barriers. For the low-frequency regime, and for critical frequency marking the onset of the conductivity's dispersion, the Barton–Nakajima–Namikawa relationship is fulfilled.

12
Polyampholyte Gels and Networks

A number of studies on the volume phase transition of anionic, cationic and non-ionic hydrogels has been reviewed [151,152]. Little attention, however, has been paid to amphoteric polyelectrolyte hydrogels [153,154]. In contrast to uniform polyelectrolytes the swelling degree of polyampholytes depends on the concentration of the constituent ionic groups or the pH of the external solution [155–159]. The pH response for the secondary structure of amphoteric copolypeptide hydrogels consisting of L-glutamic acid, L-leucine, and L-lysine was investigated through *in situ* monitoring using a Fourier transform IR microscope [160]. The hydrogels were found to exist mainly in the α-helical conformation over the whole pH range studied. Depending on the composition and pH, α-helical conformation was weakened due to the dissociation of their side chains. Amphoteric gels show a minimal swelling degree when the amount of anionic and cationic monomers is equal as in the case of linear polyampholytes considered in previous sections. As the concentration of the anionic and cationic parts deviates from the equimolar ones the swelling ratio increases rapidly (Fig. 34). A reasonable explanation of these results is that the number of osmotically active ions in the hydrogel phase increases as the molar ratio of cationic to anionic groups diverges from unity according to Donnan equilibrium, while, as the mo-

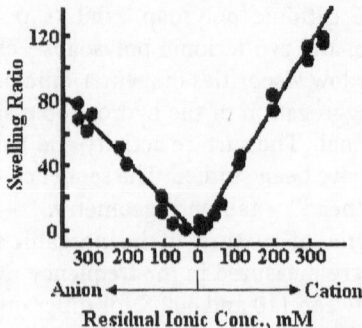

Fig. 34. Plots of swelling ratio (V/V_0) of amphoteric gel MAPTAC–SA as a function of the effective ionic density. The effective ionic density is represented by the residual amount of anionic or cationic groups after a specific binding is formed between the anionic and cationic groups. The numbers of the abscissa indicate an effective ionic density derived by subtracting the anionic group content from the cationic group content [155]

lar ratio of cationic to anionic groups approaches unity, the excess free counterions that are not needed to satisfy the electroneutrality of the chain are effectively "dialyzed" from the hydrogel interior.

Figure 35 combines the temperature-dependent and acetone-content-dependent behaviour of amphoteric gel APTAC–SA [155]. Equimolar polyampholyte gel undergoes a gradual volume change with an increase in acetone composition in an acetone/water mixture. When the content of the cationic groups increases, collapse of gel is observed. On the other hand, the same sample immersed in acetone/water mixtures of 60 and 65% acetone swells in a low

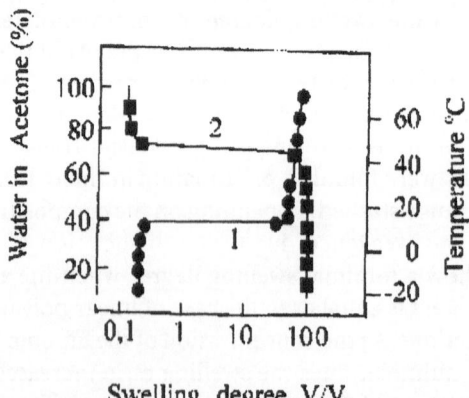

Fig. 35. Volume phase transition of amphoteric gel MAPTAC–SA in relation to temperature (*curve 1*) and acetone content (*curve 2*) [155]

temperature region and collapses in a high temperature region. While the gel in mixtures of 70% acetone or above is observed to maintain a collapsed state over a wide temperature interval, it is concluded that, although the amphoteric gel consists of various proportions of cationic and anionic groups, it exhibits the same pattern of volume phase transition, as did monoionic gels. However, the swelling behaviour observed for equimolar amphoteric gel, e.g. its continuous character, can be attributed to the specific bonding of cationic and anionic parts that act as additional cross-linking agents. It should be noted that the hydrogels containing excess cationic monomers swell more than the hydrogel containing an excess of anionic groups [154]. The "asymmetry" in swelling behaviour can be prescribed to three possible mechanisms: 1) differences in cross-link density caused by different rates of copolymerization; 2) hydrophobicity of moieties, for instance, in the case of styrene moieties that can reduce chain expansion inside the network owing to the hydrophobic interactions; and 3) differences in the activities of counterions to the charged groups, for instance, the attraction between counterions and functional groups will be weaker for MAPTAC or APTAC than for SSS or SA.

Figure 36 shows the deswelling of amphoteric gels as a function of NaCl concentration. Except for the equimolar sample that shows a low sensitivity to ionic strength all other hydrogels with excess of anionic or cationic monomers exhibit a decrease of swelling capacity with the increase of NaCl concentration. The excess of anionic or cationic moieties will require the flow of small ions inside the hydrogel to maintain the electroneutrality and, as a result, the concentration of mobile ions inside and outside of the gel becomes similar; the net osmotic pressure falls and swelling decreases.

The temperature dependence of the swelling ratios of two N-isopropylacrylamide (NIPA) based amphoteric gels was measured because the NIPA gel itself

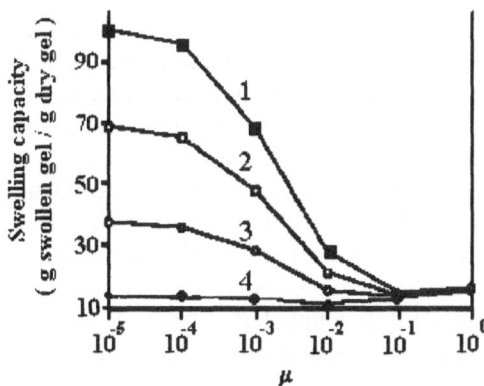

Fig. 36. Swelling equilibria for MAPTAC–SSS amphoteric hydrogels as a function of NaCl concentration. [MAPTAC]/[SSS]=7:1 (*curve 1*), 3.3:1 (*curve 2*),1.8:1 (*curve 3*), 1:1 (*curve 4*) [156]

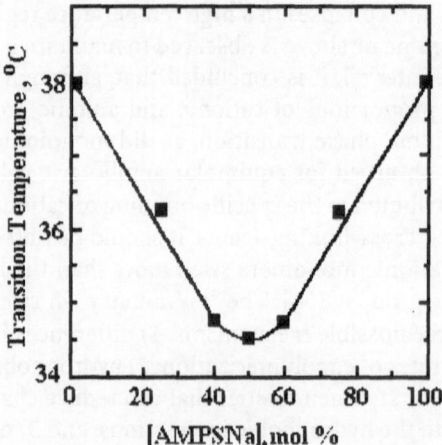

Fig. 37. Plots of the transition temperatures of the NIPA/NDAPD/AMPSNa amphoteric gels against the AMPSNa concentration. The total density of the amphoteric gels are fixed at 1 mol% (Reproduced with permission from [161])

shows a thermosensitive character in water by changing the temperature [161,162]. Figure 37 shows the equilibrium swelling curves of the amphoteric gels in water together with those of NIPA gel. When the content of anionic and cationic groups is equal in NIPA-based gel its swelling behaviour is similar to NIPA gel itself. When the ratio of cationic and anionic groups deviates from 50:50 the transition temperature and the swelling ratio increase. These results were explained by considering that the ionic contacts formed between the cationic and anionic units affect the swelling pressure of the gel. It was also found that the NIPA/betaine amphoteric gels containing an equal constituent in the pendant chain exhibit a similar swelling behaviour to integral type polyampholytes. Amphoteric hydrogel films obtained by subjecting the solution mixture of polyvinylalcohol, polyacrylic acid and polyallylamine hydrochloride to repetitive freezing and thawing swell both in acidic and basic solutions and shrink in neutral pH regions because of polyanion–polycation association [163].

Synthetic polymer gels are known to exist in two phases, swollen and collapsed. Polyampholyte gels should show a re-entrant volume transition in response to pH [164]. Tanaka et al. [165] have reported the existence of more than two phases in randomly distributed polyampholytes of MAPTAC and AA. Let us consider the behaviour of a sample containing 65 mol% of AA and 35 mol% of MAPTAC. At neutral pH this sample has a diameter $d=d_0$ that is referred to as phase 1 ($d/d_0=1$) according to the label in the original paper. At pH=8.5 the gel swells discontinuously to phase 2.7 (Fig. 38). If the pH is lowered from 8.5 to 7.0, the gel returns to phase 1 discontinuously. If instead the pH is increased further from 8.5 to 9.8 and goes back from this point the gel collapses into phase 1 at pH=6.4. The same experiments were done from the acidic side. ^{13}C NMR spectra

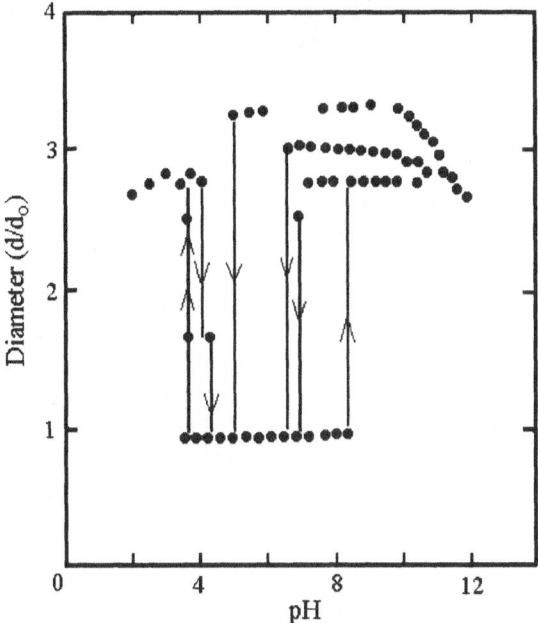

Fig. 38. Equilibrium swelling degree of d/d_0 of AA–MAPTAC (35:65 mol%) copolymer gel in water as a function of pH at 298 K [165]

for different phases at the same pH are clearly distinguishable, indicating that each phase has a different local environment. The nature of multiple phases may be understood if the competition among various interactions: mixing enthalpy and entropy, Coulomb interactions (between acidic and basic groups of polyampholytes, between charged groups and counterions), hydrogen bonding etc. is taken into account. However, more extensive study is needed to identify the microscopic structure of the new phases. Recently, a theoretical description of multiple phases in ionic copolymer gels was presented [166]. The equations for the dissociation equilibrium of the fixed charges in the gel are solved together with a Flory-type swelling model which incorporates both polyelectrolyte and polyampholyte effects. This model correctly predicts the qualitative shape of the gel swelling vs. pH curve.

Polyampholyte microgels METMAC–SAMPS have been synthesized by using an inverse microemulsion route [167]. The swelling properties of so-called "balanced" microgel particles containing variable amounts of the cross-linker were studied by photon-correlation spectroscopy as a function of the ionic strength. The microgel particles flocculate below a certain electrolyte concentration. Above this concentration they are stable and the size of the particles appears to be insensitive to both ionic strength and cross-linker concentration in the monomer feed. Transmission-electron spectroscopy results provided in pure water show that samples consist of both aggregates and isolated particles. The latter are somewhat ellipsoidal in the collapsed state. The swelling behaviour of "unbalanced" polyampholyte microgel particles displayed only a very small decrease in size with an increase of the ionic strength.

The interaction of high charge density polyampholyte gels formed from sodium methacrylate and N,N-dimethyl-N,N-diallylammonium bromide with both ionic surfactants – cetyltpyridinium chloride, sodium dodecylbenzenesulfonate – and organic dyes – alizarin, ninhydrin – has been studied [168]. The efficiency of adsorption of surfactants and dyes leading to the collapse of the gel is determined by the ratio of charges of the polyampholyte network and organic ion concentration.

It is well known that polyelectrolyte gels swell, shrink or bend when DC electric current is applied [169]. These properties of gels are applicable for the construction of chemomechanical devices, artificial muscles, energy conversion systems etc. [170]. Osada and co-workers [171] have constructed an eel-like gel actuator on the basis of poly(2-acrylamido-2-methylpropane sulfonic acid) and studied its chemomechanical properties. Polyampholyte gel is bent to the cathode or anode side if it has predominantly negative or positive charges along the macromolecules (Fig. 39) [172]. As seen from Fig. 39, the amplitude of deflection is gradually decreased with the approach to the IEP. This is probably due to

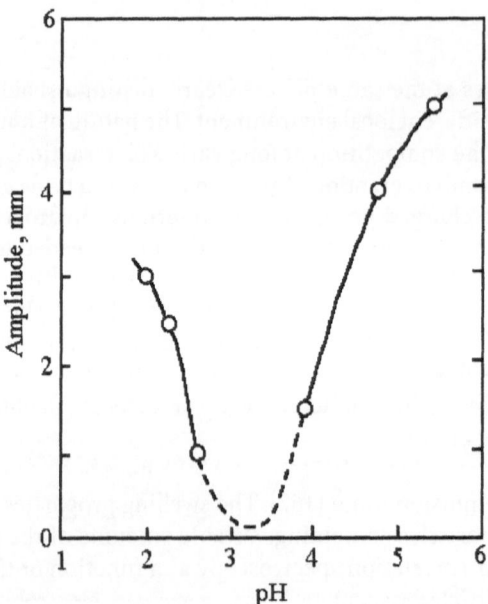

Fig. 39. Dependence of amplitude of deflection on pH of the external solution for polyampholyte gel of the vinyl ether of monoethanolamine–acrylic acid (VEMEA–AA) with $pH_{IEP}=4.2$

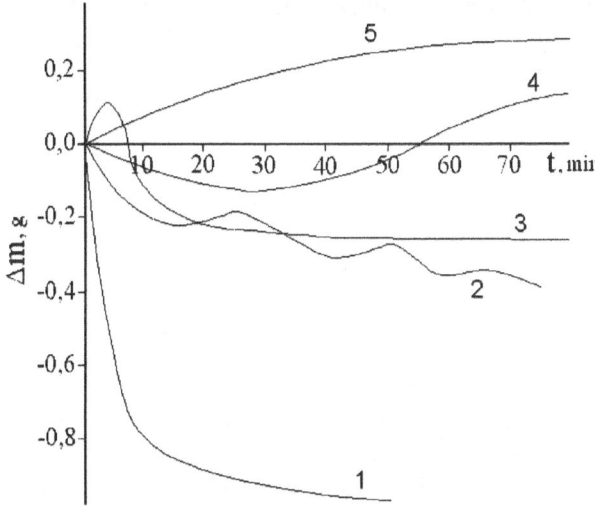

Fig. 40. Relative mass change of polyampholyte gels of the vinyl ether of monoeth-anolamine–acrylic acid (VEMEA–AA) under the influence of DC electric current at pH=3.0 (*curve 1*), 4.2 (IEP) (*curves 2,4,5*), 6.2. (*curve 3*) at $\mu=1\cdot10^{-2}$ (*curves 1–3*), $1.5\cdot10^{-2}$ (*curve 4*) and $2.5\cdot10^{-2}$ M NaCl (*curve 5*)

the formation of intraionic contacts that decrease the net charge of amphoteric macromolecules. Figure 40 represents the change in polyampholyte weight in relation to time under the influence of DC electric current [173,174].

In cationic (pH=3.0) and anionic (pH=6.2) forms, the hydrogel gradually shrinks from both ends as schematically shown below.

Under the action of DC electric field the redistribution in the double electric layer occurs and shrinking of the specimen from both side takes place. It is interesting to note that the coexistence of three phases (triphasic equilibrium), i.e.

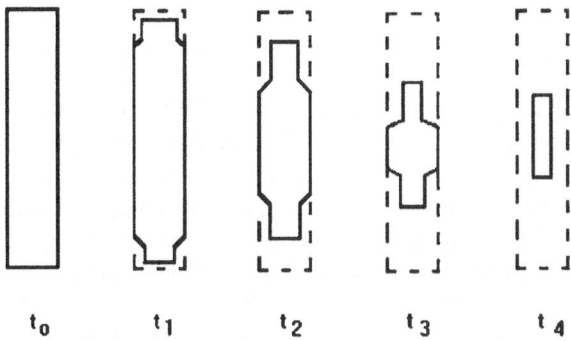

Scheme g

swollen gel, shrunken gel and pure solvent, the so called "bottleneck" macro-scopic phase boundary, was observed for ionized NIPA gel rod on heating [175]. Ampholytic gel specimens swell at the IEP (Fig. 40, curve 5). This phenomenon can be interpreted as follows: at the IEP the neutral network under the action of DC is polarized. As a result at the ends of the specimen adjoining the cathode and anode sides the density of the network charge is increased. The increased charges should be compensated by outer counterions that, in turn, cause a water flow inside the system and swelling. It was also observed that, in relation to the ionic strength, the polyampholyte gel oscillates [176]. At the IEP, at intermediate values of ionic strength $2 \cdot 10^{-2} < \mu < 7.5 \cdot 10^{-3}$, polyampholyte gel oscillates (Fig. 40, curves 2 and 4), while anionic and cationic polyelectrolytes only swell or shrink under the same conditions. Faraday law in differential form [176] suc-cessfully describes the relaxation-oscillation curves. Gel systems demonstrating rhythmically pulsatile motion similar to that of a heart beat have been developed [177]. Self-oscillations of swelling and deswelling for the poly(N-isopropylacry-lamide-co-acrylic acid) hydrogels were realized by coupling pH and temperature in the external solution media. The pH oscillations were also observed for the system consisting of iodate-sulfite-thiosulfate and poly(2-acrylamido-2-methyl-1-propanesulfonic acid) when polymeric acid was substituted for sulfuric acid [178]. Recently, an analytically tractable model for the description of polyam-pholytes in an external electric field has been presented [179]. The equilibrium properties of polyampholytes in strong electric fields decisively depend on the charge distribution along the chain.

13
Application of Polyampholytes

Water-soluble and water-swelling polyampholytes are used in a wide number of applications including sewage treatment, flocculation, enhanced oil recovery etc. The desalination of water by cross-linked polyampholytes can be regulated by changing the temperature. Such polyampholytes are called thermoregenera-ble resins (TRR). To perform the thermoregeneration the next equilibrium should take place in dependence of temperature:

$$R-COOH + R^*-NR_2' + K^+A^- \Leftrightarrow R-COO^-K^+ + R^*-N^+HR_2'A^-$$

where R and R* are polymer chains; COOH and NR$_2'$ are functional groups; K$^+$ and A$^-$ are cation and anion, respectively; and R' is a hydrogen or alkyl group.

In principle the function of TRR is as follows: salts absorbed at room temper-ature (e.g. NaCl) can easily be regenerated by hot water, e.g. the exchange equi-librium is shifted at room temperature to the right and at high temperature to the left. At room temperature the proton is transferred from the acid to the base forming charged ion-exchanging zones (COOH$\rightarrow$NR$_2'\rightarrow$COO$^-$N$^+$HR$_2'$). The heating of water from 298 K to 358 K leads to the accumulation of H$^+$ and OH$^-$ due to the ionization of water molecules; the concentration of H$^+$ and OH$^-$ increases approximately 30 times. Hydrogen and hydroxyl ions suppress the degree of ionization of the amphoteric resin and the equilibrium shifts to the left side.

The salt-rejecting properties in dynamic reverse osmosis membranes made from P2VP-*co*-PMAA were studied by Stille et al. [180,181] The desalination ability of blockpolyampholytes is minimal at the IEP. Cross-linked pentablock-polyampholytes comprising quaternary ammonium residues, styrenesulfonate and neutral cross-linkable isoprene blocks are also applicable as membranes for desalination [182]. An amphoteric ion-exchange membrane was also prepared by chemical grafting of AA and DMAEM on ozonized polyethylene and its physico-chemical properties were compared with those of monofunctional membranes [183]. The amphoteric membrane was characterized by measuring ion-exchange capacity, water and electrolyte content, electrical conductivity, membrane potential and transport number. The ion-exchange capacity of AA–DMAEM membrane determined at various pH is reported in Table 18. Curves of the electrical resistance R as a function of the external pH plotted in Fig. 41 show that the maximal R probably corresponds to the IEP and the value of R is similar for anionic exchange membrane at low pH and cationic exchange membrane at high pH. Potentiometric behaviour of amphoteric membrane prepared from a mixture of the amphoteric chitosan and poly(vinyl alcohol) was studied [184].

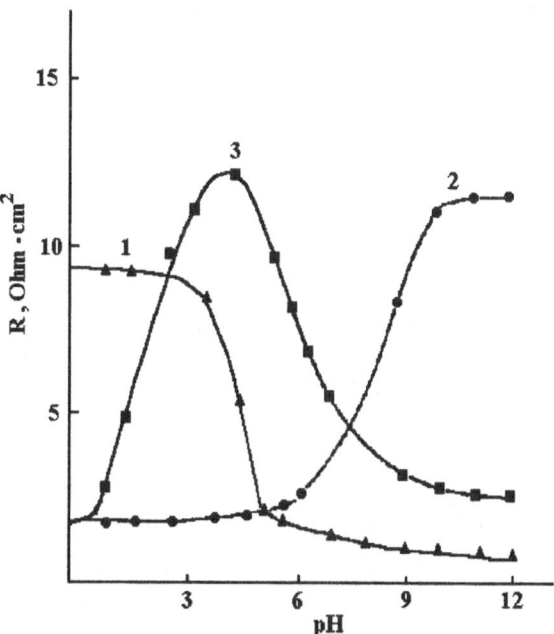

Fig. 41. Variation of the electrical resistance of cationic exchange (*curve 1*), anionic exchange (*curve 2*) and amphoteric exchange (*curve 3*) membranes as a function of the external pH (Reproduced with permission from [183])

Table 18. Ion-Exchange Capacity of Amphoteric Membranes

Labeled Ion	Cl^-	Na^+	Cl^-	Na^+
External pH	2	6	6	12
Ion-exchange capacity ($meq \cdot G^{-1}$)	0.87	0.24	0.39	0.65

The ability of polyampholyte networks to regulate the water permeability has been shown [185]. The maximal permeability corresponds to the IEP.

Several types of copolymers of methacrylic acid with various quaternary ammonium monomers, terpolymers of these copolymers with methylmethacrylate, were tested for water clarification and decolorization [186]. The polymer is an adjunct to a magnetic iron oxide employed as the primary coagulant, which is recycled after alkaline treatment. Copolymers containing phenolic or amino acid moieties as well as quaternary ammonium sites proved to be the most effective in enhancing water treatment performance with concomitant facile regeneration behaviour.

Protein purification by selective phase separation with anionic and cationic polyelectrolytes has been widely discussed by Dubin et al. [187]. The driving force of such a process is the intermolecular association with the participation of Coulomb interactions, hydrogen bonds and hydrophobic forces. Phase separation with polyampholytes has been described [188,189]. Due to the self-precipitation of polyampholytes at their own IEP in the absence of proteins, they can be used for protein extraction and purification.

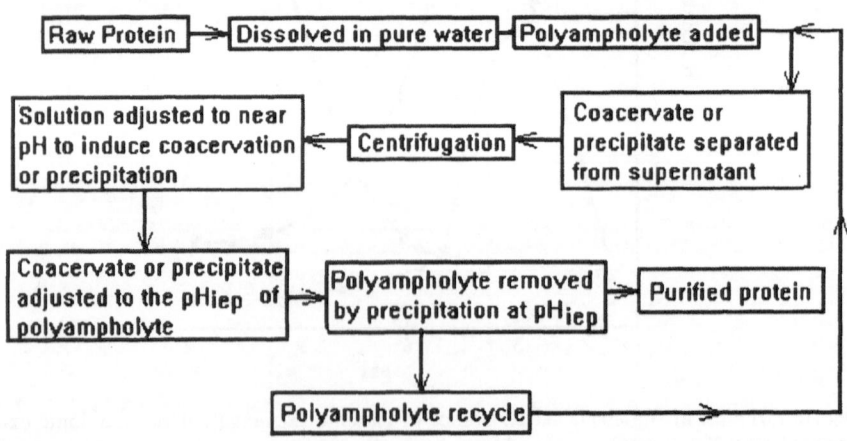

Scheme h

The main advantage of polyampholytes related to the protein separation process is the realization of the "forcing out" effect at the IEP of polyampholytes which will facilitate polymer recycling after protein separation.

The drag reduction behaviour of three classes of polyampholytes having high charge density (copolymers containing NaAMPS and AMPTAC), low charge density (terpolymers containing acrylamide and less than 15 mol% NaAMPS and AMPTAC) as well as betaine copolymers with sulfonate moieties have been examined [190,191]. A dependence of drag reduction effectiveness on copolymer structure, composition and solvation was observed. All polyampholytes exhibit higher drag reduction with increasing ionic strength of the solution. The betaine copolymers have the best drag reduction properties, the low charge density polyampholytes exhibit intermediate drag reduction effectiveness and the high charge density polyampholytes the poorest.

Polyampholyte microcapsule membranes of poly(L-lysine-*alt*-terephthalic acid) were used as self-regulating drug delivery systems [192]. The rate of permeation of 5-sulfosalicylic acid and phenyltrimethylammonium chloride through microcapsule membranes has been found to change depending on the pH and ionic strength of the medium. The drastic change in permeability lies between pH=4 and 6. A remarkable change in microcapsule size was observed in the same way as in the permeability of membranes. At pH >6 (μ=0.0154), the size of the microcapsule was about 60 μm, while it decreased to about 35 μm at pH=4 and increased again to about 45 μm when the pH decreased from 4 to 2.

The formation of coordination and ionic bonds between functional groups of VEMEA–AA and metal ions is accompanied by the gradual shrinking of polyampholyte gel [193]. During the complexation process a shell layer is formed on the surface and moves to the core region of the gel as schematically illustrated below:

It should be noted that the concentration of absorbed copper reaches a constant value after 2 h while the volume contraction of gel continues to change for up to 10 h. This is probably due to the rearrangement of polyampholyte–copper complexes. The complexed layer will probably retard the further penetration of copper ions deep in the gel and the complex formation kinetics are diffusion limited. Figure 42 shows the kinetics of deswelling of polyampholyte gel and desorption of Cu^{2+} at various pH. The desorption of Cu^{2+} from the inside of the gel is accompanied by swelling of the gel. The degree of recovery of copper ions at the IEP of polyampholytes reaches up to 60–70%.

Scheme i

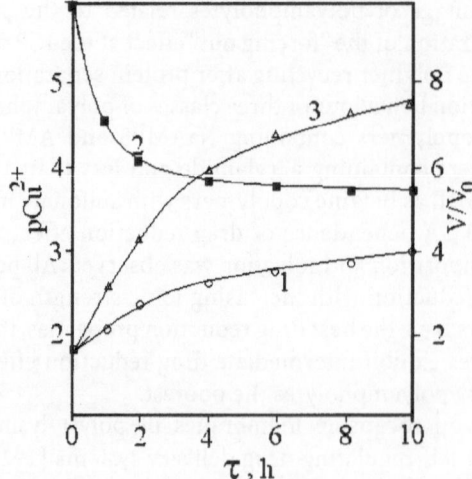

Fig. 42. Time dependence of swelling degree ratio (V/V_0) and concentration of Cu^{2+} at pH= 5.7 (*curve 1*), 5.2 (*curve 2*) and 4.8 (*curve 3*)

14
Concluding Remarks

Polyampholytes have unique electrochemical, hydrodynamic, conformational and complex-forming properties due to the presence of both acidic and basic groups. The ability of linear and cross-linked amphoteric macromolecules to adopt globular, coil, helix and stretched conformations and to demonstrate coil-globule, helix-coil conformational transitions, and sol-gel, collapsed-expanded phase or volume changes in relation to internal (nature and distribution of acid and base substituents, copolymer composition, molecular weight distribution, hydrophobicity etc.) and external (pH, temperature, ionic strength, organic solvents addition etc.) factors, as well as to form superstructures will constantly attract the attention of theorists and experimentalists.

The structures of amphoteric macromolecules, especially at the IEP, fall within eyeshot of several disciplines, at least polymer and colloid sciences, and molecular biology. Here, the ideas, approaches, theoretical calculations and experimental observations of de Gennes [194], Lifshitz et al. [195], Frenkel [196,197], Joanny [96], Burchard [198], Ise [56] and Khokhlov [199] seem to be useful to make some conclusions. According to de Gennes [194], the behaviour of macromolecules is mainly determined by topological and geometrical properties of systems. It is also suggested [195] that the existence of a high-ordered structure inside of globular particles – "Configurational information" or "linear memory" of macromolecular chains – determines the properties of materials [197]. Many similarities exist between covalently bound structures and macromolecular associates [198]. The high probability of self-organization processes in polyelectrolytes is due to the effective competition between different interaction types [199]. At the IEP, macromolecular chains form a globular structure stabilized by cooperative ionic contacts, hydrogen bonds and hydrophobic interactions, the state of which are close to hydrated globular proteins. In turn, such globular structures can be organized into various types of associates or aggregates stabilized mostly by physical forces. Random association of globules can produce colloid particles. Regular assemblies of globular particles can lead to the appearance of supercrystalline states. The accumulated data imply that polyampholytes can repeat, more or less, the structural organizations of proteins. Having recognized these facts, attempts could be directed to find out the molecular organization levels of amphoteric macromolecules.

The overall chain conformation of polyampholytes with a balanced (same number of negative and positive monomers) and nonbalanced (an excess of positive or negative charges) stoichiometry is essentially controlled by attractive (polyampholyte effect) and repulsive (polyelectrolyte effect) electrostatic interactions. Therefore, depending on the ionic strength, the solution behaviour is dominated by either polyampholyte or polyelectrolyte effects. Polyampholytes with balanced stoichiometry are insoluble in pure water but become soluble upon addition of 1:1 electrolytes owing to the screening of opposite charges and unfolding of the macromolecules, while nonstoichiometric polyampholytes are water-soluble, possess extended polyelectrolyte conformation but have the tendency to be insoluble at a high ionic strength of the solution. An "antipolyelectrolyte" character gives the stoichiometric polyampholytes good swelling in high salinity media and enables them to be used for desalination and enhanced oil recovery.

Analysis of the literature data shows that considerable progress has been made in theoretical considerations in spite of some discrepancies between theory and experiments. In future the quantum-chemical and computer simulations of polyampholyte chains will be developed.

The competition between intra- and interionic contacts is the driving force behind the behaviour that causes the "forcing out" effect near the IEP. This phenomenon can successfully be used in matrix polymerization, and for the recovery and purification of both low- and high-molecular-weight substances. Systematic structural investigations on zwitterionic "polysoaps" that combine the advantages of polyampholytes and micelles are in progress.

A study of complex formation properties of polyampholytes with respect to metal ions, dyes, organic probes, detergents and polyelectrolytes can be of help, on the one hand, in understanding the mechanism of protein denaturation, enzymatic reactions, replication, and, on the other hand, in using such processes in coagulation, purification and enrichment technologies etc.

Evidently, polyampholyte gels belong to multiphase systems due to the ability of macromolecules to adopt different stable conformations in response to changes in environmental conditions. Therefore, more extensive study is needed to identify the microscopic structure of multiphases. Stimuli of the responsive character of linear and cross-linked polyampholytes can be of help to construct thermoreversible and semipermeable membranes, microcapsules for drugs etc. The development of the oscillating behaviour of gels that will be useful for the design of "muscular" or periodic drug delivery systems, actuators and other devices has just started. According to Tanaka [200] polyampholyte gels are similar to p-n junction diodes. The flow of counterions in such gels, corresponding to the flow of holes and electrons in a diode, would be curious to examine.

Acknowledgments. The author is thankful to Professor H.Hocker for encouragement. This work was supported by Grant number INTAS-KZ-95-31.

15
References

1. Tanford Ch (1965) Physical Chemistry of Macromolecules (in Russian). Mir, Moscow
2. Bekturov EA, Kudaibergenov SE (1996) Catalysis by Polymers. Hüthig & Wepf Verlag Zug, Heidelberg
3. Bekturov EA, Bakauova ZKh (1986) Synthetic Water Soluble Polymers in Solutions. Huthig & Wepf, Heidelberg
4. Bekturov EA, Kudaibergenov SE, Rafikov SR (1990) J.Macromol.Sci., Rev. Macromol.Chem.Phys. C30: 233
5. Bekturov EA, Kudaibergenov SE, Rafikov SR (1991) Uspekhi Khimii 60: 835
6. Edwards SF, King PR, Pincus P (1980) Ferroelectrics 30: 3
7. Higgs PG, Joanny J-F (1991) J.Chem.Phys. 94: 1543
8. Kantor Y, Kardar M (1991) Europhys. Lett. 14: 421
9. Kantor Y, Li H, Kardar M (1992) Phys.Rev.Lett. 69: 61; Kantor Y, Kardar M, Li H (1994) Phys.Rev.A: 49:1383
10. Gutin A, Shakhnovich E (1994) Phys.Rev. 50: R3322
11. Dobrynin AV, Rubinstein M (1995) J.Phys. II France 5: 677
12. Qian C, Kholodenko AL (1988) J.Chem.Phys. 89: 5273
13. Schiessel H, Oshanin G, Blumen A (1996) Macromol. Theory Simul. 5: 45; Schiessel H, Blumen A (1997) Macromol. Theory Simul. 6:103; Schiessel H, Sokolov IM, Blumen A (1997) Phys.Rev.A: 56: 2390
14. Srivastava D, Muthukumar M (1996) Macromolecules 29: 2324; Muthukumar M (1996) J.Chem.Phys. 104: 691
15. Bratko D, Chakraborty AK (1996) J. Phys. Chem. 100: 1164
16. Kudaibergenov SE, Shayakhmetov ShSh, Zhaimina GM, Bekturov EA, Rafikov SR (1983) Dokl.Akad.Nauk SSSR 273: 1161; Sigitov VB, Kudaibergenov SE, Bekturov EA, Rafikov SR (1986) Dokl.Akad.Nauk SSSR 291: 403
17. Mazur J, Silberberg A, Katchalsky A (1959) J.Polym.Sci. 35: 43
18. Merle Y, Merle-Aurby L (1982) Macromolecules 15: 360
19. Merle Y (1985) J. Chim.Phys. 82: 653; Merle Y (1987) J.Phys.Chem. 91: 3092
20. Katchalsky A, Gillis J (1949) Recl.Trav.Chim.Pays-Bas. 68: 871
21. Katchalsky A, Miller IR (1954) J.Polym.Sci. 13: 57
22. Kruglova NA, Savinova IV (1979) Vysokomol. Soedin., SerA: 21: 282
23. Mosalova LF, Kruglova NA, Vorontzov ED, Evdakov VP (1982) Vysokomol. Soedin., SerA: 24: 176
24. Harris FE, Rice SA (1955) J.Phys.Chem. 58: 725; Rice SA, Harris FE (1956) J.Chem.Phys. 24: 326
25. Koper GJM, Borkovec M (1996) J. Chem. Phys. 104: 4204

26. Patrickios CS, Hertler WR, Abbot NL, Hatton TA (1994) Macromolecules 27: 930
27. Pirogov VS, Dmitrienko LV, Kipper AI, Samsonov GV (1972) Zh.Prikl.Khim. 65: 626
28. Mitra RP, Atreyi M, Gupta RG (1967) J.Electroanal.Chem. 15: 399
29. Mitra RP, Atreyi M, Gupta RG (1968) J.Electroanal.Chem. 18: 227
30. Guo-Zhen Zheng, Meshitsuka G, Ishizu A (1994) Polymer International 34: 241
31. Tanchuk YuV, Yablonko BM (1984) Ukr.Khim.Zh. 50: 88
32. Tanchuk YuV, Yablonko BM (1988) Ukr.Khim.Zh. 54: 1099
33. Tanchuk YuV, Kotenko SI, Yablonko BM (1988) Ukr.Khim.Zh. 54: 990
34. Kudaibergenov SE, Shayakhmetov ShSh, Bekturov EA (1979) Dokl.Akad.Nauk SSR 246: 141; Izv.Akad.Nauk KazSSR, Ser.Khim 3: 67
35. Bekturov EA, Kudaibergenov SE, Shayakhmetov ShSh (1980) Vysokomol.Soedin. Ser.B: 22: 91
36. Bekturov EA, Kudaibergenov SE (1986) J.Macromol.Sci.,Phys. B25: 133
37. Guo-Zhen Zheng, Meshitsuka G, Ishizu A (1995) J.Polym.Sci: Part B: Polym. Phys. 33: 867
38. Lubina SYa, Strelina IA, Sogomonyantz ZhS, Dmitrieva SI, Korotkina OZ, Tarasova VS, Skazka VS, Yamshikov VM (1970) Vysokomol.Soedin. Ser.A: 12: 1560
39. McCormick CL, Johnson CB (1988) Macromolecules 21: 686, 694
40. Kathman EE, Salazar LC, McCormick CL (1991) Polymer Preprints 32: 98
41. McCormick CL, Salazar LC (1992) Macromolecules 25: 1896
42. McCormick CL, Salazar LC (1992) Polymer 33: 4384
43. Corpart JM, Candau F (1993) Colloid Polym.Sci. 271: 1055; Neyret S, Baudouin A, Corpart JM, Candau F (1994) Biophysics 16: 669
44. Corpart JM, Selb J, Candau F (1993) Polymer 34: 3873; Ohlemacher A, Candau F, Munch JP, Candau SJ (1996) J.Polym.Sci.Polym.Phys.Ed. 34: 2747
45. Corpart JM, Candau F (1993) Macromolecules 26: 1333
46. Skouri M, Munch JP, Candau SJ, Neyret S, Candau F (1994) Macromolecules 27: 69
47. Everaers R, Johner A, Joanny JF (1997) Europhys.Lett. 37: 275
48. Varoqui R, Tran Q, Pefferkorn E (1979) Macromolecules 12: 831
49. Bekturov EA, Kudaibergenov SE, Frolova VA, Schulz RC, Zoller J (1990) Makromol.Chem. 191: 457; Kudaibergenov SE, Frolova VA, Bekturov EA, Rafikov SR (1990) Dokl.Akad.Nauk SSSR 311: 296
50. Bazt MR, Skorikova EE, Vikhoreva LS, Gal'braikh LS (1990) Vysokomol.Soed. Ser A: 32: 805
51. Waldo A-M, Carlos P-C (1993) Makromol.Chem.,Rapid Commun. 14: 73
52. Patrickios CS, Hertler WR, Hatton TA (1995) Fluid Phase Equilibria 108: 243
53. Bekturov EA, Kudaibergenov SE, Khamzamulina RE, Nurgalieva DE, Schulz RC, Zoller J (1992) Makromol.Chem., Rapid Commun. 13: 225
53a.Giebeler E, Stadler R (1997) Macromol. Chem. Phys. 198: 3816
54. Kudaibergenov SE, unpublished results
55. Ise N (1996) Ber. Bunsen-Ges. Phys.Chem. 100: 841
56. Ise N, Matsuoka H, Ito K (1989) Polym.Prepr. 30: 333
57. Sackman E (1994) Macromol.Chem.Phys. 195: 7
58. Finkenaur AL, Dickinson L, Charles CJ (1984) Amer.Chem.Soc.Polym.Prepr. 25: 109
59. Chiso Y-Ch, Strauss UP (1985) Amer.Chem.Soc.Polym.Prepr. 26: 214
60. Zheltonozhskaya TB, Pop GS, Eremenko BV, Uskov IA (1981) Vysokomol.Soedin., Ser.A: 21: 2425
61. Kang SK, Jhon MS (1993) Macromolecules 26: 171
62. Yang JH, Jhon MS (1995) J.Polym.Sci: Part A: Polym.Chem. 33: 2613
63. Kurmaeva AI, Vedikhina LI, Sverdlov LB, Barabanov VP (1985) Kolloid.Zh. 47: 1185
64. Vedikhina LI, Kurmaeva AI, Barabanov VP (1985) Vysokomol.Soedin., Ser.A: 27: 2131
65. Kurmaeva AI, Vedikhina LI, Barabanov VP (1988) Kolloid.Zh. 50: 1011, 1189
66. Barabanov VP, Kurmaeva AI, Vedikhina LI, Avvakumova NI (1985) Vysokomol.Soedin., Ser.A: 27: 2543

67. Hartley FR, Burgess C, Alkok RM (1980) Solution Equilibria. Ellis Horwood, New York
68. Kudaibergenov SE, Zhaimina GM, Sigitov VB, Bekturov EA (1985) Izv.Akad.Nauk KazSSR, Ser.Khim: 1: 34
69. Bekturov EA, Kudaibergenov SE (1987) J.Macromol.Sci.,Phys. B26: 75
70. Wittmer J, Johner A, Joanny JF (1995) J.Phys. II 5: 635
71. Furukawa J, Kobayashi E, Doi T (1979) J.Polym.Sci., Part A: Polym.Chem. 17: 255
72. Smets G, Samyn C, (1979) In: Selegny E (ed) Optically Active Polymers. Reidel, Dordrecht, Holland
73. Bekturov EA, Kudaibergenov SE, Sigitov VB (1986) Polymer 27: 1269; (1987) Koord.Khim. 13: 600
74. Sigitov VB, Kudaibergenov SE, Vozzhova NA, Andrusenko AA, Shaikhutdinov EM, Bekturov EA (1986) Izv.Akad.Nauk KazSSR, Ser.Khim: 4: 40
75. Ormanova PS, Abilov ZhA, Musabekov KB (1984) Vysokomol.Soedin., Ser.B: 26: 506
76. Bekturov EA, Kudaibergenov SE, Kanapyanova GS (1983) Makromol.Chem. Rapid Commun. 4: 653
77. Bekturov EA, Kudaibergenov SE, Kanapyanova GS (1984) Polym.Bull. 11: 551
78. Bekturov EA, Kudaibergenov SE, Kanapyanova GS (1984) Kolloid.Zh. 46: 861
79. Bekturov EA, Kudaibergenov SE, Khamzamulina RE, Schulz RC, Zoller J (1991) Makromol.Chem. Rapid Commun. 12: 37
80. Savinova IV, Fedoseeva NA, Evdakov VP, Kabanov VA (1976) Vysokomol. Soedin., Ser A: 18: 2050
81. Skorikova EE, Vikhoreva GA, Kalyuzhnaya RI, Zezin AB, Galbraikh LS, Kabanov VA (1988) Vysokomol. Soedin., Ser A: 30, 44
82. Koetz J, Hahn M, Philipp B, (1989) Acta Polym. 40: 405; Kudaibergenov SE, Sigitov VB, Bekturov EA, Koetz J, Philipp B (1989) Vysokomol.Soedin. Ser.B: 31: 132; Koetz J, Philipp B, Kudaibergenov SE, Bekturov EA (1991) Acta Polym. 42: 181
83. Kudaibergenov SE, Bekturov EA, Rafikov SR (1990) Dokl.Akad.Nauk SSSR 311: 296
84. Bekturov EA, Kudaibergenov SE, Frolova VA, Khamzamulina RE, Schulz RC, Zoller J (1990) Makromol.Chem. 191: 457
85. Tultaev AK, Kudaibergenov SE, Bekturov EA (1990) Izv.Akad.Nauk KazSSR, Ser.Khim: 6: 37
86. Morawetz H, Hughes WL (1952) J.Phys.Chem. 56: 64
87. Nath S (1995) J.Chem.Techn.Biotechn. 62: 295; Nath S, Patrickios CS, Hatton TA (1995) Biotechnol. Prog. 11: 99; Patrickios CS, Hertler WR, Hatton TA (1994) Biotechnology and Bioengineering 44: 1031
88. Petrov RV, Khaitov RM, Norimov ASh, Vinogradov IV, Kabanov VA, Mustafaev MI (1982) Immunology 6: 52
89. Takahashi A, Kawaguchi M (1982) Adv.Polym.Sci. 46: 1
90. Norde W, Lyklema J (1978) J.Colloid & Interface Sci. 66: 266, 277, 285
91. Shirahama H, Shikuma T, Suzawa T (1989) Colloid & Polym.Sci. 267: 587
92. Serra J, Puig J, Martin A, Galisteo F (1992) Colloid & Polym.Sci. 270: 574
93. Kurmaeva AI, Yusupova RI, Barabanov VP, Osipova EN, Vedikhina LI, Mahyurov IR, Potapova MV (1991) Kolloid. Zh. 53: 866
94. Blaackmeer J (1990) PhD thesis, University of Wageningen, The Netherlands
95. Ouali L, Neyret S, Candau F, Pefferkorn E (1996) J. Colloid and Interface Sci. 76:86
96. Joanny J-F (1994) J.Phys.II France 4: 1281; Dobrynin AV, Rubinstein M, Joanny JF (1997) Macromolecules 30: 4332
97. Tusupbaev NK, Musabekov KB, Kudaibergenov SE (1998) Macromol.Chem.Phys. 199: 401
98. Baran AA (1986) Polymer-containing disperse systems, Naukova Dumka, Kiev
99. Ainakulova ZhM, Tusupbaev NK, Musabekov KB (1994) Izv.Natl.Akad. Nauk Republic of Kazakstan, Ser.Khim: 4: 47
100. Derjaguin BV, Landau LD (1941, 1945) Acta Phys.Chim. 11: 802; 14: 633; 15: 663; Verwey EJ, Oberbeek JThG (1948) Theory of the stability of lyophobic colloids, Elsevier, Amsterdam, New York

101. Gregory J (1988) Colloids and Surfaces 31: 231
102. Kudaibergenov SE, Bekturov EA (1989) Vysokomol.Soedin. Ser.A: 31: 2614
103. Bekturov EA, Kudaibergenov SE (1989) Makromol.Chem.Macromol.Symp. 26: 281
104. Kudaibergenov SE (1991) Doctoral thesis, Moscow State University, Moscow, Russia
105. Jellinek HHG, Luh MD (1969) J.Polym.Sci., Part A-1: 7: 2445
106. Koetz J, Philipp B, Kudaibergenov SE, Sigitov VB, Bekturov EA (1988) Colloid. Polym.Sci. 266: 906
107. Koetz J, Philipp B, Kudaibergenov SE, Sigitov VB, Bekturov EA (1989) Acta Polym. 40: 405
108. Kudaibergenov SE, Sigitov VB, Bekturov EA, Rafikov SR, Koetz J, Philipp B (1989) Izv.Akad.Nauk KazSSR 6: 33
109. Koetz J, Hahn M, Philipp B, Bekturov EA, Kudaibergenov SE (1993) Makromol. Chem. 194: 397
110. Kulagina EM (1995) PhD thesis, Kazan State Technological University, Kazan, Russia
111. Salamone JC, Tsai CC, Olson AP, Watterson AC (1978) Amer.Chem.Soc.Polym.Prepr. 19: 261
112. Salamone JC, Tsai CC, Watterson AC, Olson AP (1978) Polymer 19: 1157
113. Salamone JC, Tsai CC, Olson AP, Watterson AC (1980) In: Eisenberg A (ed) Ionic polymers. ACS Series 187: 22
114. Salamone JC, Rodrigues EA, Lin KC, Quach L, Watterson AC (1980) Polymer 26: 1241
115. Salamone JC, Watterson AC, Hsu TD, Tsai CC, Mahmud MU, Wisniewski AW, Israel SC (1978) J.Polym.Sci.Polym.Symp. 64: 229
116. Salamone JC, Watterson AC, Hsu TD, Tsai CC, Mahmud MU (1977) J.Polym.Sci. Polym.Lett.Ed. 15: 487
117. Salamone JC, Tsai CC, Watterson AC, Olson AP (1980) In: Goethals EJ (ed) Polymeric amines and ammonium salts. Pergamon Press, Oxford, p 105
118. Salamone JC, Tsai CC, Olson AP, Watterson AC (1980) J.Polym.Sci.Polym.Chem.Ed. 18: 2983
119. Salamone JC, Ahmed I, Elayaperumal P, Rahja MK, Watterson AC, Olson AP (1986) Polym.Mater.Sci.Eng. 55: 269
120. Watterson AC, Salamone JC, Elayaperumal P (1986) Polym.Prepr. 27: 275; Salamone JC (1988) J.Macromol.Sci.Chem. 25: 811
121. Salamone JC, Rice WC Encyclopedia of polymer science and engineering,2nd ed., Wiley-Interscience, New York, 1988, vol.11, p 514
122. Hart P, Timmerman D (1958) J.Polym.Sci. 28: 638
123. Itoh Y, Abe K, Saburo S (1986) Makromol.Chem. 187: 1691
124. Schulz DN, Peiffer DG, Agarwal PK, Larabee J, Kaladas J,J, Soni L, Handwerker B, Garner RT (1986) Polymer 27: 1734
125. Liaw DJ, Lee WF, Whung YC, Lin MC (1987) J.Appl.Polym.Sci. 34: 999; Liaw DJ, Huang CC (1996) Polymer Intern. 41: 267; Liaw DJ, Huang CC, Chou YP (1997) Eur.Polym.J. 33: 829
126. Wielema TA, Engberts JBFN (1988) Eur.Polym.J. 24: 647
127. Wielema TA, Engberts JBFN (1990) Eur.Polym.J. 26: 639
128. Cardoso J, Manero O (1991) J.Polym.Sci.Polym.Phys.Ed. 29: 639
129. Ouchi T, Nomoto K, Hosaka Y, Imoto M, Nakaya T, Iwamoto T (1984) J.Macromol.Sci. Chem. A21: 859
130. Nakaya T, Toyoda H, Imoto M (1986) Polym.J. 18: 881
131. Muroga Y, Amano M, Katagiri A, Noda I, Nakaya T (1995) Polym.J. 27: 65
132. Kato T, Takahashi A (1996) Ber.Bunsenges.Phys.Chem. 100: 784
133. Lee W-F, Tsai C-C (1995) Polymer 36: 357
134. Monroy Sato VM, Galin JC (1984) Polymer 25: 121
135. Monroy Sato VM, Galin JC (1984) Polymer 25: 254
136. Galin M, Marchal E, Mathis A, Meurer B, Monroy Sato VM, Galin JC (1987) Polymer 28: 1937

137. Zheng YL, Knoesel R, Galin JC (1987) Polymer 28: 2297
138. Galin JC, Galin M (1992) J.Polym.Sci: Part B: Polym.Phys. 30: 1103
139. Galin M, Galin JC (1993) Makromol.Chem. 194: 3479
140. Galin JC, Galin M (1992) J.Polym.Sci: Part B: Polym.Phys. 30: 1113
141. Galin JC, Galin M (1992) J.Polym.Sci: Part B: Polym.Phys. 30: 1103
142. Galin M, Chapoton A, Galin JC (1993) J.Chem.Soc.,Perkin Trans. 2: 545
143. Huglin MB, Radwan MA (1991) Polymer International 26: 97
144. Laschewsky A, Zerbe I (1991) Polymer 32: 2081
145. Anton P, Laschewsky A (1991) Makromol.Chem.Rapid Commun. 12: 189
146. Koberle P, Laschewsky A, Lomax TD (1991) Makromol.Chem.Rapid Commun. 12: 427
147. Laschewsky A (1991) Colloid.Polym.Sci. 269: 785
148. Anton P, Koberle P, Laschewsky A (1993) Makromol.Chem. 194: 1
149. Anton P, Laschewsky A (1993) Makromol.Chem. 194: 601
150. Rozanski SA, Kremer F, Koberle P, Laschewsky A (1995) Makromol.Chem.Phys. 196: 877
151. Adv.Polym.Sci. (1992) 109: 1
152. Adv.Polym.Sci. (1993) 110: 1
153. Starodubtzev SG, Ryabina VR, Khokhlov AR (1987) Vysokomol.Soedin., Ser.A: 29: 2281
154. Myoga A, Katayama S (1987) Polym.Prepr.Japan 36: 2852; Myoga A, Katayama S (1988) Kobunshi Kagaku 37: 530, 757
155. Katayama S, Myoga A, Akahori Y (1992) J.Phys.Chem. 96: 4698
156. Baker JP, Stephens DR, Blanch HW, Prausnitz JM (1992) Macromolecules 25: 1955; Baker JP, Blanch HW, Prausnitz JM (1995) Polymer 36: 1061
157. Wen S, Stevenson WTK (1993) Colloid & Polym.Sci. 271: 38
158. Kudaibergenov SE, Nurgalieva DE, Bekturov EA (1994) Macromol.Chem.Phys. 195: 3033
159. Kudaibergenov SE, Nurkeeva ZS, Mun GA, Sigitov VB (1995) Reports Natl.Acad.Sci. Republic of Kazakhstan 2: 58
160. Kugo K, Nakanishi E, Hasegawa M, Nishino J (1995) Kobunshi Ronbunshu 52: 461
161. Wada N, Yagi Y, Inomata H, Saito S. (1993) J.Polym.Sci. Polym.Chem.Ed. 31: 2647
162. Yu H, Grainger DW (1993) Polymer Preprints 34: 829
163. Suzuki M, Hirasa O. (1993) Adv.Polym.Sci. 110: 241
164. Hirokawa Y, Tanaka T, Sato E (1985) Macromolecules 18: 2784
165. Annaka M, Tanaka T (1992) Nature 365, 430; Shibayama M, Tanaka T (1993) Adv. Polym.Sci. 109: 1
166. Mafe S, Manzanares JA, English AE, Tanaka T (1997) Phys.Rev.Lett. 79: 3086; English AE, Mafe S, Manzanares JA, Grosberg Ayu, Tanaka T (1996) J. Chem. Phys. 104: 8713; English AE, Tanaka T, Edelman ER (1998) Macromolecules 31: 1989
167. Neyret S, Vincent B (1997) Polymer 38: 6129
168. E.E.Makhaeva, Le Thi Minh Thanh, Abstr. 1st Intern.Symp. on Polyelectrolytes, September 18–22, 1995, Potsdam, Germany
169. Shiga T, Kurauchi T (1990) J.Appl.Polym.Sci. 39: 2305
170. Osada Y (1987) Adv.Polym.Sci. 82: 3
171. Ueoka Y, Isogai N, Nitta T, Okuzaki H, Gong JP, Osada Y (1994) Rreprints of Sapporo Symposium on Intelligent Polymer Gels, p 117
172. Kudaibergenov SE (1996) Ber. Bunsen-Ges. Phys.Chem. 100: 1079
173. Suleimenov IE, Sigitov VB, Salina AA, Kudaibergenov SE (1998) Vysokomol. Soedin. Ser:A: 40: 5
174. Sigitov VB, Salina AA, Kudaibergenov SE (1996) Vestnik KazGU 5: 203
175. Hirotsu S (1993) Adv.Polym.Sci. 110: 1
176. Kudaibergenov SE, Sigitov VB, Suleimenov IE, Nurkeeva ZS Reports of Russian Acad. Sci. (1998) (to be published)
177. Yoshida R, Ichijo H, Hakuta T, Yamaguchi T (1995) Macromol.Rapid Commun. 16: 305

178. Giannos SA, Dinh SM, Brener B (1995) Macromol.Rapid Commun. 16: 527
179. Winkler RG, Reineker P (1997) J.Chem.Phys. 106: 2841
180. Kamachi M, Kurihara M, Stille JK (1972) Macromolecules 5: 161
181. Kurihara M, Kamachi M, Stille JK (1973) J.Polym.Sci., Part A: Polym.Chem. 11: 587
182. Itou H, Toda M, Ohkoshi K, Iwata M, Fujimoto T, Miyaki Y, Kataoka T (1988) Ind.Eng.Chem.Res. 27: 983
183. Elmidaoui A, Bouyevin B, Belcadi S, Gavach C (1991) J.Polym.Sci., Part B: Polym. Phys. 29: 705
184. Saito K, Tanioka A (1996) Polymer 27: 5117
185. Bekturov EA, Kudaibergenov SE (1988) Vestnik Acad.Nauk KazSSR 12: 41
186. Anderson NJ, Bolto BA, Eldridge RJ, Jackson MB (1993) Reactive Polym. 19: 87
187. Xia J, Dubin PL (1994) Protein-polyelectrolyte complexes. In: Dubin P, Block J, Davis R, Schultz DN, Thies C (eds) Macromolecular complexes in chemistry and biology. Springer, Berlin Heidelberg New York, p 247
188. Patrickios CS, Hertler WR, Hatton TA (1994) In: Schmitz KS (ed) Macro-ion characterization: from dilute solution to complex fluids. ACS, Washington DC, chap 19
189. Patrickios C, Jang CT, Hertler WR, Hatton AT (1993) Polymer Preprints 34: 954
190. Mumick PS, Welch PM, Salazar LC, McCormick CL (1994) Macromolecules 27: 323
191. Mumick PS, Welch PM, Hester RD, McCormick CL (1992) Polymer Preprints 33: 337
192. Makino K, Miyauchi E, Togawa Y, Ohshima H, Kondo T (1992) Polymer Preprints 33: 476
193. Sigitov VB, Koblanov SM, Ospanova GSh, Kudaibergenov SE, Nurkeeva ZS (1997) Reports of Natl.Acad.Sci. Republic of Kazakhstan 2: 72
194. de Gennes PG (1979) Scaling concepts in polymer physics. Cornell University Press
195. Lifshitz IM, Grosberg AYu, Khokhlov AR (1979) Uspekhi Fiz.Nauk 127: 353
196. Frenkel SYa (1973) Polymers: problems, prespectives, prognosis In: Physics for today and tomorrow. Nauka, Leningrad, p 176
197. Bartenev GM, Frenkel SYa (1990) Physics of polymers. (in Russian), Khimiya, Leningrad, p 432
198 Burchard W (1993) Trends in Polym.Sci. 1: 192
199. Khokhlov AR, Dormidontova EE (1997) Usp.Fiz.Nauk (Adv.Phys.Sci.) 167: 113
200. Katayama S, Hirokawa Y, Tanaka T (1984) Macromolecules 17: 2641

Editor: Prof. H. Höcker
Received: April 1998

Author Index Volumes 101–144

Author Index Volumes 1–100 see Volume 100

Subject Index

Springer
and the
environment

At Springer we firmly believe that an international science publisher has a special obligation to the environment, and our corporate policies consistently reflect this conviction.
We also expect our business partners – paper mills, printers, packaging manufacturers, etc. – to commit themselves to using materials and production processes that do not harm the environment. The paper in this book is made from low- or no-chlorine pulp and is acid free, in conformance with international standards for paper permanency.

 Springer